D1520725

WITHDRAWN

SPECTROSCOPIC ELLIPSOMETRY AND REFLECTOMETRY

SPECTROSCOPIC ELLIPSOMETRY AND REFLECTOMETRY

A USER'S GUIDE

Harland G. Tompkins
Motorola, Inc.

William A. McGahan
Nanometrics, Inc.

A WILEY-INTERSCIENCE PUBLICATION

JOHN WILEY & SONS, INC.

New York • Chichester • Weinheim • Brisbane • Singapore • Toronto

This book is printed on acid-free paper. ♾

Published simultaneously in Canada.

Library of Congress Cataloging-in-Publication Data:

Tomkins, Harland G.
Spectroscopic ellipsometry and reflectometry : a user's guide / by Harland G. Tompkins, William A. McGahan.
p. cm.
"A Wiley-Interscience publication."
Includes bibliographical references and index.
ISBN 0-471-18172-2 (cloth)
1. Ellipsometry. 2. Reflectometer. 3. Materials–Optical properties. 4. Thin films–Optical properties. 5. Surfaces (Technology) I. McGahan, William A., 1966– . II. Title.
QC443.T63 1999
681'.25—dc21 98–38199

Printed in the United States of America.

10 9 8 7 6 5 4 3 2 1

CONTENTS

PROTOTYPICAL ANALYSES

PREFACE

Ellipsometry and reflectometry are currently being used extensively in industry to measure the thickness and other properties of thin layers of dielectrics and semiconductors and, to a small extent, conductors. The techniques have been used in scientific laboratories for over a century. The use of these techniques in industry is more recent, starting in the 1960s and becoming more prevalent in the 1980s.

Both of the techniques are used for metrology, where the sample structure is well understood and only the film thickness is required. The increased speed of reflectometry and the smaller spot size makes it a desirable tool for the large number of measurements required for statistical process control on a patterned wafer. Both techniques are also used for analysis, where the structure (and optical constants) are often unknown. The additional information provided by ellipsometry (two measured quantities) makes this technique particularly desirable for analysis. Ellipsometry is also preferable for measuring very thin layers, that is, submonolayer coverages to several tens of atoms deep.

There are several commercial suppliers of instruments for both techniques. Instruments which are intended as metrology tools (as opposed to analytical tools) often have extensive material handling robotics and pattern recognition equipment. The cost of the metrology part of the tool is often only a small part of the total cost, which is dominated by the material handling part. The subject matter of this book is the measurement and analytical part of ellipsometry and reflectometry; hence we will not deal with material handling aspects.

Although there is a vast amount of material on ellipsometry and reflectometry in the technical literature, there are few textbooks. The classic book by Azzam and Bashara[1] was published in 1977 and is very useful for designing ellipsometers. This work also contains extensive theory. *A User's Guide to Ellipsometry* by H. G. Tompkins[2] was published in 1993 on the subject of single-wavelength ellipsometry. This work was intended for beginners who had a working single-wavelength ellipsometer and wanted either to understand how it worked or to stretch its use beyond the turnkey applications.

By the mid-1990s, the use of spectroscopic ellipsometry increased significantly. It seemed appropriate that a beginner-level book be written on this subject. By its nature, reflectometry is also a spectroscopic technique and it might be observed that there is no textbook readily available on reflectometry. Several manufacturers are now marketing tools which use both ellipsometers and reflectometers. Accordingly, this book is intended for the casual user of these two techniques who wants to know how the techniques work, how to optimize their use, and how to stretch the techniques beyond the turnkey application.

Both ellipsometry and reflectometry are spectroscopic optical techniques. After a brief introduction, we begin with the fundamentals of polarized light and ellipsometry. One of the distinguishing features of spectroscopic ellipsometry, and to some extent of spectroscopic reflectometry, is the structure of the optical constant spectrum. Chapter 3 takes the fundamentals a bit farther, discussing why the optical constants of materials are the way that they are. If one is satisfied to accept optical constants in whatever form one finds them, Chapter 3 can be skipped until a later date. Chapter 4 deals with the instrumental aspects of reflectometers, explaining how the reflectometers and ellipsometers do what they do. This is not intended as an alternative to reading the instrument manufacturers manual, but to give it perspective. Chapters 5 and 7 discuss the look of reflectance and ellipsometric spectra, pointing out the salient features and discussing why they look the way they do. It is the opinion of the authors that one must understand single-wavelength ellipsometry before one can understand spectroscopic ellipsometry well; hence Chapter 6 gives the salient features of single-wavelength ellipsometry. In Chapters 8 and 9, we discuss the analytical approach for collecting and analyzing ellipsometric and reflectance data, and in Chapter 10 we discuss how one assures oneself that the instrument is operating properly. This is followed by a chapter each on very thin films and roughness. Following these 12 chapters, we have a series of prototypical analyses. Here we suggest a step-by-step approach to several typical analytical situations. Finally, there are several appendices for the reader who wishes to acquire some additional background information.

Finally, no one writes a book such as this alone. First, we would like to acknowledge the support of our wives, Rose Ann Tompkins and Jen McGahan, who provided emotional support while we spent hours in front of a computer screen. Second, we would like to acknowledge the support of our employers, managers, and co-workers. Without this encouragement, writing a book such as this would be very difficult. In addition, we would like to acknowledge the many useful discussions with Mark Keefer, KLA-Tencor. In our attempt to provide balance to the book, one of us (HGT) sought his contributions and he cheerfully provided much information, particularly on the reflectometry technique.

Tempe, AZ
Sunnyvale, CA
October 1998

HARLAND G. TOMPKINS
WILLIAM A. MCGAHAN

1. R. M. A. Azzam and N. M. Bashara, *Ellipsometry and Polarized Light*, North Holland, Amsterdam, 1977.
2. H. G. Tompkins, *A User's Guide to Ellipsometry*, Academic Press, New York, 1993.

SPECTROSCOPIC ELLIPSOMETRY AND REFLECTOMETRY

CHAPTER 1

Perspective and History

1.1 HISTORICAL ASPECTS

The concept of using light to measure the thickness of a thin film is quite old. Brewster[1] attributes the first observation of colors from "thin bubbles of essential oils, spirit of wine, turpentine, and soap and water" to Robert Boyle. Sir Isaac Newton observed interference rings when a convex lens was placed on a flat piece of glass, and by observing the colors along with the separation produced a table which, according to Brewster,[2] "are of extensive use and may be regarded as presenting us with a micrometer for measuring minute thicknesses of transparent bodies by their colors, when all other methods would be inapplicable."

These observations are based on interference phenomena and would presently be called *interferometry* or *reflectometry*. Although Malus observed polarized light[3,4] in 1810, it was another 80 years before the phase difference between two mutually perpendicular polarized light beams was used by Drude[5] to measure the thickness of very thin films. This principle is the basis of the method which was later called *ellipsometry*.[6]

Polarized light will be discussed at length in a later chapter. Although the most familiar form is linearly polarized light, the most general form is elliptically polarized light. The technique of ellipsometry takes its name from this aspect of light. Aspnes[7] has observed that the shape of the ellipse is used in ellipsometry and the overall size of the ellipse is used in reflectometry; hence when the beam is totally polarized, the combination of ellipsometry, reflectometry, and transmission use all the information which is carried by the light beam.

Although reflectometry has been used since the time of Sir Isaac Newton in academic and research laboratories, it began to show up in industrial laboratories as a home-built bench-top characterization tool in the 1960s.[8] Commercial instruments appeared in the late 1970s. Nanometrics marketed a microscope-based instrument with manual positioning. The thickness was determined by the position of the maxima and minima in the reflectance spectra, with variations in the refractive index being ignored. The analytical algorithm was relatively simple. Since that time, companies such as Nanometrics, Prometrix, and Tencor (and combinations thereof) have added analytical features such as the use of optical constant tabular data and dispersion relationships, curve fitting, regression analysis, and so on. Wafer handling improvements include computer controlled positioning, mapping, pattern recognition, etc. Other improvements include smaller spot sizes and extended wavelength ranges, particularly into the UV.

The first report of the determination of the thickness of thin films with what is now called ellipsometry was by P. Drude.[5] In his case, only very thin films were observed. Alexandre Rothen[6] introduced the name ellipsometry in 1945. The human eyeball was the detector used to determine when a *null* had been reached. The development of photomultiplier detectors greatly increased the accuracy of the method. In the early years, monochromatic light sources were usually mercury or sodium arc lamps. The development of the laser again stimulated the development of the technique. The mainframe computer allowed the extensive use of regression analysis and significantly improved the speed of calculation. The transition to minicomputers and then to desktop computers also added the possibility of control of the equipment in addition to simply calculating the results. All of these technological developments aided in the evolution of ellipsometry. Probably the most significant stimulus, however, was the development of planar technology in the microelectronics industry. Archer[9] recognized that this technique could readily be used to measure the thicknesses of the oxides of silicon, and he was instrumental in implementing the method into this industry.

At the present time, any microelectronics fab will have several reflectometry instruments and at least one single-wavelength ellipsometer. Spectroscopic ellipsometers have been used in government laboratories and in some large industrial laboratories for about 15 years. Since the early 1990s, spectroscopic ellipsometers began to appear in the research and development centers of many companies as well as in a few microelectronics fabs. A few manufacturers are beginning to market instruments which have both a spectroscopic ellipsometer and a reflectometry instrument.

1.2 COMPLEMENTARY NATURE OF THE TECHNIQUES

The primary strength of single-wavelength ellipsometry (SWE) is its ability to measure very thin layers. If the material is well understood, SWE can readily be used to measure layers which are a few tens of angstroms thick. For in situ experiments, Archer,[10] in the early 1960s, showed that this technique can measure a fraction of a monolayer of adsorbed gas.

For layers which are well understood and a few hundreds to a few thousands of angstroms thick, reflectometry has the advantage of speed. In addition, since reflectometry is usually done at normal incidence, a much smaller feature can be measured. Reflectometry does not suffer from the periodicity problem which makes SWE a bit difficult for films which are thicker than half a micrometer.

Both SWE and reflectometry excel when the sample is a single layer on a substrate. One of two layers can be analyzed if all of the information is available for the other layer. In general, these two techniques suffer when the material is not well understood (i.e. a material with unknown composition) or when multiple layers are involved.

The primary strength of spectroscopic ellipsometry (SE) is its ability to analyze multiple layers and to determine the optical constant dispersion relationship. From these optical constants, microstructural information can be deduced. Examples of this are the degree of crystallinity of annealed amorphous silicon, and the aluminum fraction in $Al_xGa_{1-x}As$. The combination of SE and reflectometry on a single instrument provide the speed needed for metrology and the ability to analyze in depth when unknown materials are involved.

1.3 FORMAT AND PURPOSE OF THE BOOK

Owing to the nature of the authors' employers, many of the examples and illustrations in this book will be taken from the microelectronics industry. These techniques are also used extensively in the recording media industry, corrosion science, and various other industries.

There was a surge of development in ellipsometry in the 1960s and 1970s which resulted in several international conferences on this subject. The proceedings[11–13] of these conferences are a rich source of information on the varied aspects of scientific research in this field. In addition, a collection of key papers[14] on ellipsometry was published in 1991 by SPIE. Azzam and Bashara[15] published a book in 1977 on the subject of ellipsometry and this book has been the key reference work on this subject since that time. In 1993, Tompkins[16] published a user's guide for single-wavelength ellipsometry. The intention of that work was to provide a guide for the casual user, and to give the essence of the technique without requiring the reader to go through all the minute details of the method.

With the emergence of spectroscopic ellipsometry, a user's guide on this subject (with the same philosophy) seemed appropriate. Reflectometry is spectroscopic in nature, is based on many of the same physical principles, and is often used for similar purposes as spectroscopic ellipsometry. We know of no comprehensive book on reflectometry as it is currently being practiced. Accordingly, we chose to combine these two techniques in a single work.

The general format of this book will be as follows. We start with the physics of light and reflections. One of the most powerful elements of these techniques is the texture (not only magnitude, but shape) of the optical constant spectra. Whereas two different transparent materials which have the same index of refraction at a given wavelength cannot be distinguished with SWE, the fact that they will not have the same values at all wavelengths allow us to distinguish them easily. An excellent example is the ability to distinguish amorphous silicon from crystalline silicon. A very important chapter discusses why the optical constants are the way they are.

The middle section's of the book deal with the instrumental aspects of these techniques and the essence of both a reflectance spectrum and an ellipsometric spectrum. It is the opinion of the authors that it is necessary to understand the essence of single-wavelength ellipsometry in order to understand spectroscopic ellipsometry. Accordingly, we include a chapter on the essentials of single-wavelength ellipsometry.

There are sections on methods, approach, data analysis, and reassuring oneself that the instrument is operating properly. Two chapters on very thin films and roughness deal with some special situations.

The latter part of the book consists of a section containing 10 prototypical analyses. This is written in a how-to-do-it manner for several materials which occur frequently in the microelectronics industry. Finally, we include several appendices for those who wish to delve a bit more deeply into some of the subjects.

1.4 REFERENCES

1. D. Brewster, *Treatise on Optics*, Longman, Rees, Orme, Brown, Green, and Taylor, London, 1831, p. 100.
2. D. Brewster, *Treatise on Optics*, Longman, Rees, Orme, Brown, Green, and Taylor, London, p. 108.
3. D. Brewster, *Treatise on Optics*, Longman, Rees, Orme, Brown, Green, and Taylor, London, p. 164.
4. D. S. Kliger, J. W. Lewis, and C. A. Randell, *Polarized Light in Optics and Spectroscopy*, Academic Press, New York 1990. p. 3 (these authors list the date as 1808).
5. P. Drude, *Annalen der Physik und Chemie*, **36**, 532 (1889). **36**, 865 (1889).
6. A. Rothen, *Rev. Sci. Instrum.*, **16**, 26 (1945).
7. D. E. Aspnes, in *Handbook of Optical Constants of Solids*, edited by E. D. Palik, Academic Press, New York, 1985, p. 90.
8. The historical information on reflectometry comes from Mark Keefer, KLA-Tencor, and is from a chapter in *Handbook of Thin Film Deposition Processes and Technology*, 2nd edition, K. Seshan editor, Noyes Publications. Westwood, NJ, 1999.
9. R. J. Archer, *J. Opt. Soc. Amer.*, **52**, 970 (1992).
10. R. J. Archer, in *Ellipsometry in the Measurement of Surfaces and Thin Films*, edited by E. Passaglia, R. R. Stromberg, and J. Kruger, National Bureau of Standards, Misc. Publ. 256, US Government Printing Office, Washington, 1964, pp. 255 ff.

11. *Ellipsometry in the Measurement of Surfaces and Thin Films*, Symposium Proceedings, edited by E. Passaglia, R. R. Stromberg, and J. Kruger, National Bureau of Standards, Misc. Publ. 256, Washington, 1964.
12. *Proceedings of the Symposium on Recent Developments in Ellipsometry*, edited by N. M. Bashara, A. B. Buckman, and A. C. Hall, North Holland, Amsterdam, 1969.
13. *Proceedings of the Third International Conference on Ellipsometry*, edited by N. M. Bashara and R. M. A. Azzam, North Holland, Amsterdam, 1976.
14. *Selected Papers on Ellipsometry*, edited by R. M. A. Azzam, SPIE Milestone Series, Vol. MS 27, SPIE, Bellingham, WA, 1991.
15. R. M. A. Azzam and N. M. Bashara, *Ellipsometry and Polarized Light*, North-Holland, Amsterdam, 1977.
16. H. G. Tompkins, *A User's Guide to Ellipsometry*, Academic Press, New York, 1993.

CHAPTER 2

Fundamentals

We are often called upon to measure samples of materials which are so small that quantities such as thickness cannot be determined by simply looking at them with the human eye. In such cases, we traditionally use a probe of some kind (electrons, ions, photons, etc.) and look at how the material of interest alters the probe. For spectroscopic ellipsometry and reflectometry, the probe is a light beam or electromagnetic wave in the UV, visible, or IR region. Electromagnetic waves and polarized light are treated in textbooks[1–3] and reference books,[4–7] on optics. In this section, we review the basic features which apply to ellipsometry and reflectometry.

2.1 DESCRIPTION OF AN ELECTROMAGNETIC WAVE

Historically, there were two schools of thought about light. One group considered light to consist of a beam of particles, and the other group considered light to consist of a wave of some sort. Sir Isaac Newton was the last great champion of the particle theory (until Albert Einstein discovered the photoelectric effect). After Newton, most scientists considered light to be a wave, although it was not at all clear what was *waving*.

The description of light as an electromagnetic wave continued to be developed in a rather fragmentary way until James Clerk Maxwell (1831–1879) provided a unifying description in a paper presented before the Royal Society in 1864. In *A Dynamical Theory of the Electromagnetic Field*, Maxwell proposed a theory which required the vibrations to be strictly transverse (perpendicular to the direction of propagation) and provided a definite connection between light and electricity. The results of this theory were expressed as four equations which are known as Maxwell's equations.

Maxwell's equations and the resulting wave equation are analogous to a differential equation which describes the forces on a body and the resulting equation which describes the motion of the body as a function of time. Maxwell's equations and the derivation of the wave equation are presented in several texts[3] and reference books.[4,5] In this introduction, we simply list the wave equation. This is treated in detail in Appendix B.

Briefly, an electromagnetic wave is a transverse wave consisting of an electric field vector and a magnetic field vector, both of whose magnitude is a function of position and time. The electric vector and the magnetic vector are mutually perpendicular and are both perpendicular to the direction of propagation. It is generally accepted that the human eye reacts to the electric vibration. The two are not independent, and specification of the electric field vector completely determines the magnetic field vector. For these reasons, and for simplicity in general, we consider only the electric field vibration.

The equation for an electromagnetic plane wave can be expressed several ways. Its purpose is to describe the electric field as a function of position and as a function of time. If we consider motion in one dimension only, the solution to the wave equation can be expressed as

$$E(z, t) = E_0 \sin\left(-\frac{2\pi}{\lambda}(z - vt) + \xi\right) \tag{2.1}$$

where E is the electric field strength at any time or place, E_0 is the maximum field strength or *amplitude* of the wave, z is the distance along the direction of travel, t is the time, v is the velocity of the wave, λ is the wavelength, and ξ is an arbitrary phase angle (which will allow us to offset one wave from another when we begin combining waves).

Figure 2.1 shows a pictorial illustration of such a wave with the electric field waving in the vertical direction. In this instance, we show the electric field as a function of position, with a constant time. For comparisons later when we discuss polarized light, we identify the places where the electric field is maximum, minimum, and zero.

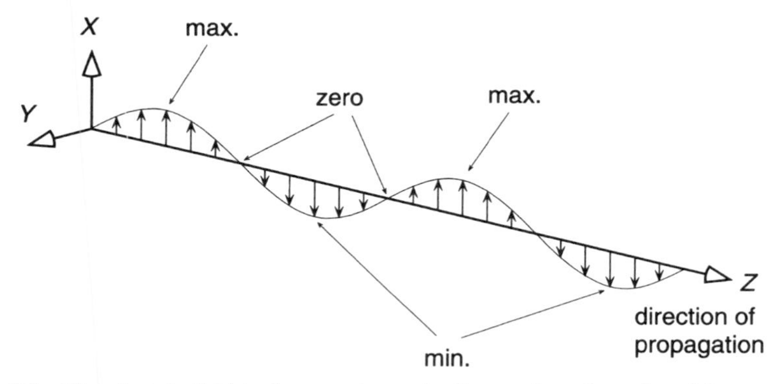

Figure 2.1 The electric field is shown schematically as a function of position, at a fixed time.

If we consider three dimensions (vector notation) and a circular function which is a bit more general than the sine function, a solution to the wave equation which comes from Maxwell's equations is

$$\bar{E}(\bar{r}, t) = \bar{E}_0 \exp\left(\frac{-\mathrm{j}2\pi\tilde{N}}{\lambda}\bar{q}\cdot\bar{r}\right)\exp(-\mathrm{j}\omega t) \tag{2.2}$$

where $\bar{r}$ is the position vector, $\bar{q}$ is the unit vector in the direction of wave propagation, $\tilde{N}$ is the complex index of refraction (to be introduced later), ω is the angular frequency, and j is the imaginary number ($\sqrt{-1}$).

We cannot measure the electric field amplitude of the electric field directly because of the high angular frequency. Instead, we measure the flux of energy of the radiation. The quantity of energy being transferred across a unit area which is perpendicular to the direction of propagation is called the *intensity* of the wave (here denoted as I). The intensity I is proportional to the square of the amplitude of the wave, that is,

$$I \propto E_0^2 \tag{2.3}$$

We note in passing that the quantity which is utilized in reflectometry is the intensity of the wave, whereas the quantity which is utilized in ellipsometry is the amplitude of the wave. Although light waves can occur in several different forms (i.e., spherical, plane, etc.), we shall deal exclusively with plane waves.

2.2 THE EFFECT OF MATTER ON ELECTROMAGNETIC WAVES

2.2.1 The Complex Index of Refraction

When a light beam (plane wave) arrives at an interface between air and another material, as depicted in Figure 2.2, several phenomena can occur. The wave generally slows down, changes direction, and in some instances begins to be absorbed. Some of the light is reflected back into the first medium (air) and does not enter the second medium. From an optics point of view, the second material is characterized by its complex index of refraction, $\tilde{N}_2$, and its thickness (in this case, taken to be infinite). In general, the complex index of refraction $\tilde{N}$ is a combination of a real number and an imaginary number and is designated as

$$\tilde{N} = n - \mathrm{j}k \tag{2.4}$$

where n is called the *index of refraction* (sometimes leading to confusion), k is called the *extinction coefficient*, and j is the imaginary number.

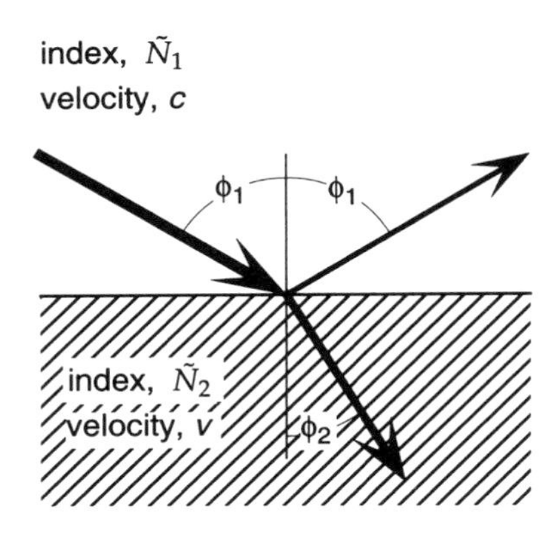

Figure 2.2 Light interacting with a plane parallel interface between air and a material with complex index of refraction $\tilde{N}_2$.

In a dielectric material such as glass, n is an inverse measure of the phase velocity of light in the material, related to the speed of light in free space, that is,

$$n = \frac{c}{v} \tag{2.5}$$

In silicon nitride, where $n \approx 2$, the phase velocity of light is half that of light in free space.

The extinction coefficient k is a measure of how rapidly the intensity decreases as the light passes through the material. In order to gain a better understanding of the extinction coefficient, let us consider first, the *absorption coefficient*, which is denoted as α and is defined as follows. In an absorbing medium, the decrease in intensity I per unit length z is proportional to the value of I. In equation form, this is

$$\frac{\mathrm{d}I(z)}{\mathrm{d}z} = -\alpha I(z) \tag{2.6}$$

The solution to this equation is

$$I(z) = I_0 e^{-\alpha z} \tag{2.7}$$

where I_0 is the intensity of the light just inside the material of interest. (α is dependent on loss of intensity due to absorption only. Loss of light due to the reflection at the interfaces does not contribute to the magnitude of α.)

This is the familiar negative exponential function which approaches zero but never gets there. The extinction coefficient k is related[8] to the absorption coefficient by

$$k = \frac{\lambda}{4\pi}\alpha \tag{2.8}$$

TABLE 2.1 Penetration Depths (Defined in Text) for Several Materials

Material	Wavelength (λ)	Extinction Coefficient (k)	Penetration Depth (D_p)
Single crystal Si[9]	6328 Å	0.016	~ 3.1 μm
Single crystal Si[10]	3009 Å	4.09	58 Å
Tungsten[11]	6328 Å	2.63	191 Å
Aluminum[11]	6328 Å	6.92	73 Å

We characterize the curve defined by Eq. 2.7 with a quantity which we will call the *penetration depth*. When the quantity αz is equal to 1.0, from Eq. 2.7, the intensity will have decreased by a factor of e^{-1} (about 37%) of its original value. We define[8] the penetration depth as the depth where this occurs, and denote it as D_p. From Eq. 2.8, this is given by

$$D_p = \frac{\lambda}{4\pi k} \tag{2.9}$$

(Note that in Eqs. 2.8 and 2.9, λ is the wavelength of light in free space rather than in the medium itself.)

We shall find that both the index of refraction and the extinction coefficient are functions of wavelength. The term *optical constants* is somewhat inappropriate. In addition, these quantities are also functions of temperature. The term *optical functions* is sometimes used. In keeping with common usage, however, we shall use the older term and, when appropriate, use the term *optical constant spectra* to denote the index of refraction and the extinction coefficient plotted as a function of either wavelength or photon energy.

Some examples of penetration depths are given in Table 2.1. In the case of single-crystal silicon, we show the value of k at two different wavelengths. Note that at a thickness of D_p, the intensity has dropped to 37% of its initial value, at $2 \times D_p$, it has dropped to 15%, at $3 \times D_p$, to 5%, and at $4 \times D_p$, to 2%. Hence, optical measurements only *see* about four penetration depths of a given material.

2.2.2. Laws of Reflection and Refraction

As suggested by Figure 2.2, some of the light is reflected at the interface and some is transmitted into the material. It was known by the ancients (Euclid, 300 BC) that the angle of reflection is equal to the angle of incidence, that is,

$$\phi_i = \phi_r \tag{2.10}$$

In Figure 2.2, both angles are listed as ϕ_1. The law of refraction is somewhat more involved and is called *Snell's law* after Willebrord Snell, who discovered the principal in 1621. Snell's law, in its most general form, is

$$\tilde{N}_1 \sin \phi_1 = \tilde{N}_2 \sin \phi_2 \tag{2.11}$$

This is derived in Appendix C.

When dealing with a dielectric material, that is, $k = 0$, the law simplifies to the more familiar

$$n_1 \sin \phi_1 = n_2 \sin \phi_2 \tag{2.12}$$

All of the terms in Eq. 2.12 are real numbers. For Eq. 2.11, generally $k = 0$ for the ambient, hence $\tilde{N}_1$ is real, and the sine function for medium 1 is a real number (as we would expect). If $\tilde{N}_2$ is a complex number (k_2 is nonzero) then the sine function in medium 2 is a complex function rather than the familiar function (opposite side over the hypotenuse).[12] There is a corresponding complex cosine function such that

$$\sin^2 \phi_2 + \cos^2 \phi_2 = 1 \tag{2.13}$$

We shall find that we use the complex cosine function in Fresnel's equations and we use Eq. 2.13 along with Snell's law to determine its value.

2.3 POLARIZED LIGHT

2.3.1 Sources

When a given photon is emitted from an incandescent source, its electric field is oriented in a given direction. The electric field of the next photon will be oriented in a different direction, and in general photons are emitted with electric fields oriented in all different directions. This is called *unpolarized* light. If we arrange for all the photons in our light beam to be oriented in a given direction, the light is referred to as *polarized* light. One familiar way to do this is to pass the light through an optical element which only allows light having one particular orientation to pass through. Another method is to have a light source which emits polarized light. Whereas incandescent light sources are unpolarized, most lasers emit light which is more or less polarized.

2.3.2 Linearly Polarized Light

In Figure 2.3, we depict the electric field strength for two light beams with the same frequency and the same amplitude traveling along the same path. We have offset them in the drawing for clarity. One is polarized in the vertical

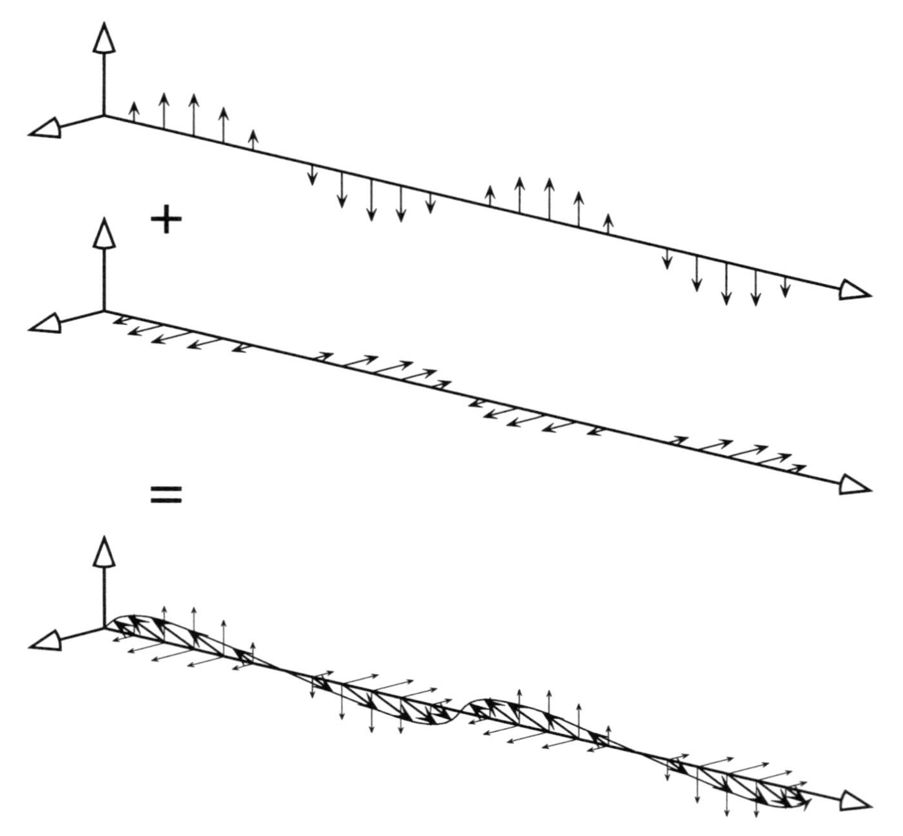

Figure 2.3 Combining two linearly polarized light beams which are in phase and have the same frequency produces linearly polarized light.

direction and one is polarized in the horizontal direction. In this case, note specifically that the maximum, minimum, and zero points of the vertical wave coincide with those of the horizontal wave, that is, the waves are *in phase*. When the vector sums of the components of the two waves are added at each point in space, the resultant wave is a linear wave which is polarized at 45° to the vertical. If all other conditions were the same, but the amplitudes were not equal, the result would have been a linear polarized wave at an angle different from 45°. Specifically, *when two linearly polarized waves with the same frequency are combined in phase, the resultant wave is linearly polarized.*

2.3.3 Elliptically Polarized Light

In Figure 2.4, we again depict two light beams with the same frequency and amplitude traveling along the same path. Again, one is polarized vertically and

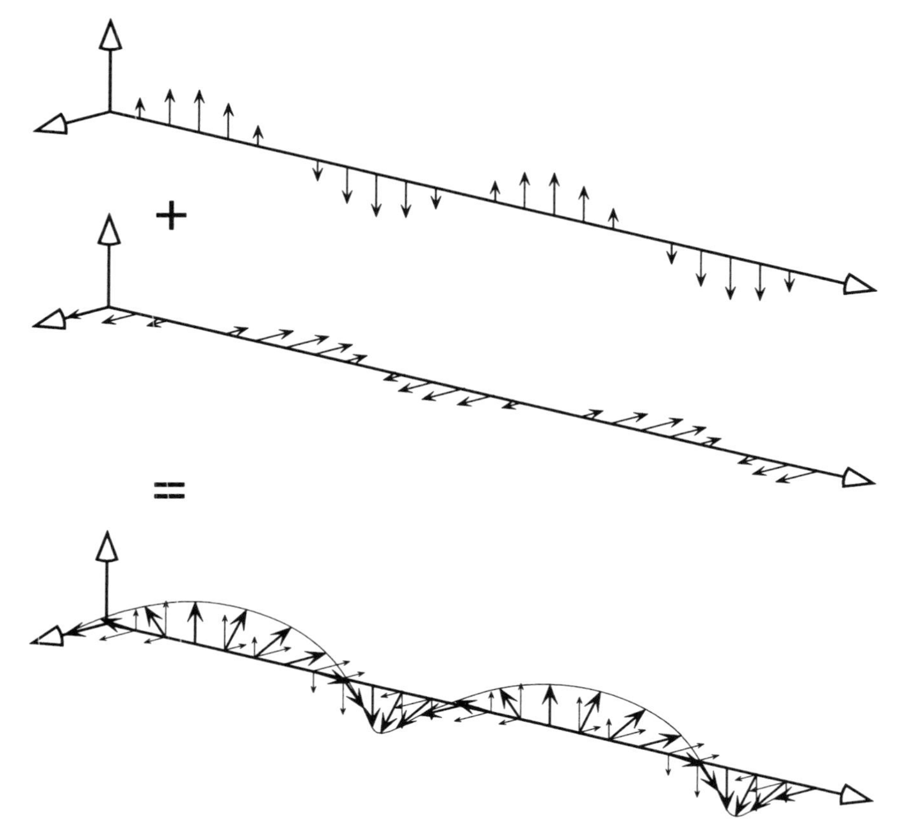

Figure 2.4 Combining two linearly polarized light beams which are a quarter wave out of phase and which have the same frequency and amplitude will produce circularly polarized light.

the other is polarized horizontally. In this case, however, the maximum, zero, and minimum of the electric field strength of the horizontal wave have been displaced from those of the vertical wave (in this particular example, the two waves are *out of phase* by 90°.) When the two waves are combined, the tips of the arrows of the resultant wave do not move back and forth in a plane, as was the case in the previous illustration. Instead, they move in a manner which, if viewed end-on, would describe a circle. This is called circularly polarized light. Had the phase shift been anything other than 90°, or had the amplitudes not been equal, the tips of the arrows would have appeared to be moving on an ellipse, if viewed end-on, and this is referred to as elliptically polarized light. Specifically, *when two linearly polarized waves with the same frequency are combined out of phase, the resultant wave is elliptically polarized.* Linearly polarized light and circularly polarized light are specific cases of the more general elliptically polarized light.

2.3.4 Application of Elliptically Polarized Light

The light used in most reflectometry instruments is not intentionally polarized. Elliptically polarized light is used in ellipsometry, and in fact is the reason for the name ellipsometry. Elliptically polarized light is generated when linearly polarized light reflects from a surface under certain conditions. The amount of ellipticity which is induced depends on the surface (optical constants, presence of films, etc.). A second method for changing the ellipticity of polarized light is to pass the light beam through certain specific optical elements. By directing our light beam at materials of interest, we use it as a probe. We use our optical elements to determine how much ellipticity was induced from the reflection and then calculate from this the properties (optical constants, thicknesses of films, etc.) of our material of interest.

2.4 THE REFLECTION OF LIGHT

2.4.1 Orientation

Both reflectometry and ellipsometry involve the light making a reflection from the surface of interest. In order to write the equations which describe the effect of the reflection on the incident light, it is necessary to define a reference plane. Figure 2.5 shows schematically a light beam reflecting from the surface. The incident beam and the direction normal to the surface define a plane which is perpendicular to the surface and this is called the *plane of incidence*. Note that the outgoing beam is also in the plane of incidence. As indicated in Figure 2.2, the angle of incidence is the angle between the light beam and the normal to the surface. The effect of the reflection depends on the polarization state of the incoming light and the angle of incidence. In Figure 2.5 we show the amplitude of the electric wave which is waving in the plane of incidence as E_p and the amplitude of the electric wave which is waving perpendicular to the plane of

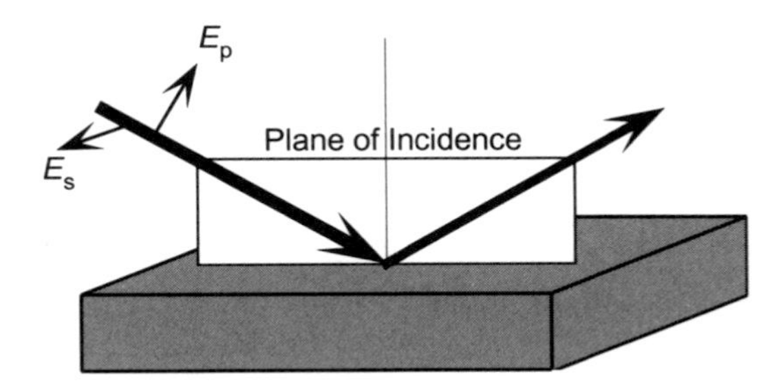

Figure 2.5 The plane of incidence is defined as the plane which contains both the incoming beam and the normal to the surface. The amplitude of the electric field wave in the plane of incidence and perpendicular to the plane of incidence are called E_p and E_s, respectively.

incidence as E_s. These waves are referred to as p-waves and s-waves, respectively. The subscripts "s" and "p" stand for the German words "senkrecht" and "parallel."

2.4.2 The Reflection Equations of Fresnel

The reflectance, denoted herein as "$\Re$", is the ratio of the *intensity* of the outgoing light compared to that of the incoming light. This is the quantity which is measured in all reflectance instruments. Recall that the intensity is proportional to the square of the amplitude of the wave.

In addition to the reflectance, we are also interested in the ratio of the *amplitude* of the outgoing wave compared to that of the incoming wave. We first focus on a single interface and will develop the more general case later. When only one interface is considered, this ratio is called the *Fresnel reflection coefficient,* and it will be different for s-waves and p-waves. As suggested by Figure 2.2, let us suppose that the interface separates medium 1 and medium 2, with respective indices of refraction $\tilde{N}_1$ and $\tilde{N}_2$ and angles of incidence and refraction ϕ_1 and ϕ_2 (related by Snell's law). When the beam is incident from medium 1 onto medium 2, the Fresnel reflection coefficients are given by

$$r_{12}^{\mathrm{p}} = \frac{\tilde{N}_2 \cos\phi_1 - \tilde{N}_1 \cos\phi_2}{\tilde{N}_2 \cos\phi_1 + \tilde{N}_1 \cos\phi_2} \qquad r_{12}^{\mathrm{s}} = \frac{\tilde{N}_1 \cos\phi_1 - \tilde{N}_2 \cos\phi_2}{\tilde{N}_1 \cos\phi_1 + \tilde{N}_2 \cos\phi_2} \tag{2.14}$$

where the superscripts refer to either p-waves or s-waves and the subscripts refer to the media which the interface separates. Corresponding equations exist for transmission. The reflection and transmission coefficient equations are derived in Appendix C.

2.4.3 The Brewster Angle

For light incident from air onto dielectrics, that is, when $k = 0$, all of the terms in the Fresnel equations above are real numbers. Figure 2.6A shows a plot of both of the Fresnel coefficients as a function of angle of incidence for a material such as TiO_2 which has an index of refraction $n = 2.2$ at a wavelength of 6328 Å. At normal incidence, r^{p} and r^{s} have equal magnitudes but opposite signs. For a single interface, the reflectance is simply the square of the Fresnel reflection coefficient. Figure 2.6B shows the reflectance, $\Re$, also plotted versus angle of incidence. At normal incidence, all of the cosine terms are equal to +1 and we have

$$r_{12}^{\mathrm{p}} = \frac{n_2 - 1}{n_2 + 1} \qquad r_{12}^{\mathrm{s}} = \frac{1 - n_2}{1 + n_2} \qquad \Re^{\mathrm{p}} = \Re^{\mathrm{s}} = \left(\frac{n_2 - 1}{n_2 + 1}\right)^2 \tag{2.15}$$

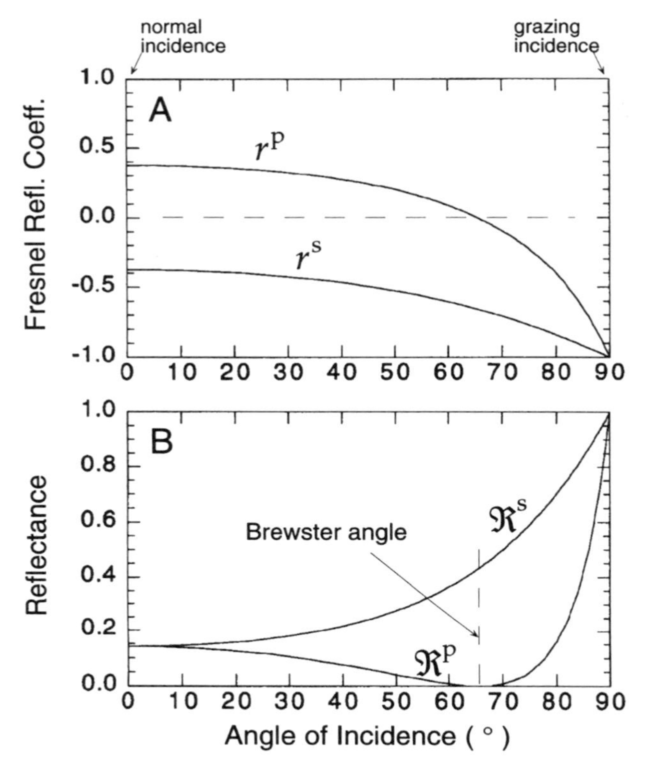

Figure 2.6 (A) The Fresnel reflection coefficients, and (B) the reflectance, plotted versus the angle of incidence for light incident from air onto a dielectric such as TiO_2 with $n = 2.2$ and $k = 0$, at a wavelength of 6328 Å. At the Brewster angle, all of the reflected light is polarized with the electric vector perpendicular to the plane of incidence.

The reflectance of the two waves must be equal at normal incidence since the plane of incidence is no longer uniquely defined.

For other than normal incidence, we see that r^s is always negative and nonzero, whereas r^p is positive for angles near-normal, passes through zero, and is negative for near-grazing angles of incidence. This can be rationalized algebraically[13] from the relationship $n_2 > n_1$ and $\cos \phi_1 > 0$.

At the angle of incidence where r^p is zero, the reflectance $\Re^p$ is also zero; hence all of the reflected light is polarized perpendicular to the plane of incidence. This is shown in Figure 2.6 as the *Brewster angle*. This phenomena was discovered by David Brewster[14] in the early 1800s. This angle is also known as the *polarizing angle* and sometimes the *principal angle*.

Two significant ramifications of the Brewster angle are that at that angle, designated as ϕ_B,

$$\tan \phi_B = \frac{n_2}{n_1} \tag{2.16}$$

and

$$\cos \phi_2 = \sin \phi_B \qquad (2.17)$$

which is to say that the angle between the reflected beam and the transmitted beam is a right angle. One additional feature of the Brewster angle for dielectrics is that this is the incidence angle where the phase shift of the p-wave on reflection shifts abruptly from zero to 180°. No such shift occurs for the s-wave.

The Brewster angle is a function of the index of refraction and, as indicated earlier, the index of refraction is a function of wavelength; hence the Brewster angle is a function of wavelength. The term *Brewster wavelength* is sometimes used with a single angle of incidence. This is simply the wavelength where the value of the index of refraction matches the Brewster condition for that angle of incidence.

The concept of the Brewster angle or polarizing angle is used routinely by photographers in photographing objects which are under water. The light coming from the underwater object (fish or alligators) is often significantly less that the light reflected from the top surface of the water, and the reflected light will obscure the underwater object. If the angle of incidence of the reflected light is roughly equal to the Brewster angle, a polarizer adjusted to the correct azimuth will remove the reflected light, allowing the camera to capture the light from the underwater object.

At the Brewster angle, although the reflected light is polarized in one direction only, the transmitted light still has components of both polarizations. Multiple interfaces can be used to remove successively more and more of the perpendicular polarized light until the transmitted light is virtually pure polarized light in the plane of incidence. Historically,[15] this is one method of obtaining polarized light.

When the reflecting surface is not a dielectric, that is, k is nonzero, the situation becomes more complicated. The Fresnel reflection coefficients r^p and r^s are now complex numbers and the concepts of greater than zero and less than zero have no meaning. Normally, there is no situation where both the real and imaginary parts of the complex number are zero; hence there is no analogous version for Figure 2.6A for metals or semiconductors. The reflectance values, $\mathfrak{R}^p$, and $\mathfrak{R}^s$, are real numbers, however, defined as the square of the magnitudes of r^p and r^s, and can be plotted, as shown in Figure 2.7, for tantalum. Although $\mathfrak{R}^p$ does not go to zero, it does go through a minimum at an angle which is called the principal angle.

As the angle of incidence increases, the phase difference between the p-wave and the s-wave shifts gradually, rather than abruptly as for a dielectric (at the Brewster angle). For metals, as for all other materials, the phase difference passes through 90° at the principal angle. This will be discussed in detail in Chapter 6 (see Figure 6.3).

It should be noted that high reflectance is obtained when the index of the substrate is significantly different from that of the ambient. This can occur

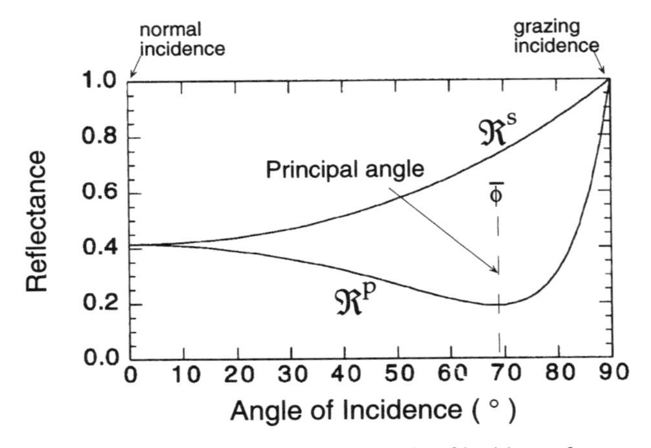

Figure 2.7 The reflectance, plotted versus the angle of incidence for a metal such as Ta with $n = 1.72$ and $k = 2.09$ at a wavelength of 6328 Å.

when n_2 is significantly different from 1.0 or when k_2 is large (significantly different from zero).

2.4.4 Reflections with Films

When more than one interface is present, that is, with a film, the light which is transmitted across the first interface cannot be ignored, as was the case in the previous section. As suggested by Figure 2.8, the resultant reflected wave returning to medium 1 will consist of light which is initially reflected from the first interface as well as light which is transmitted by the first interface, reflected from the second interface, and then transmitted by the first interface going in the reverse direction, and so on. Each successive transmission back

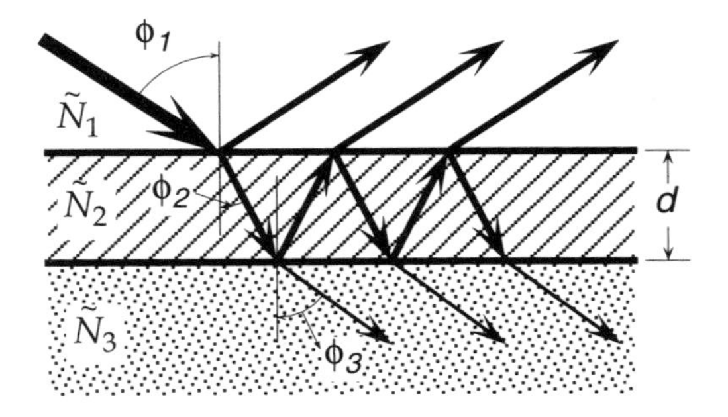

Figure 2.8 Reflections and transmissions for two interfaces. The resultant reflected beam is made up of the initially reflected beam and the infinite series of beams which are transmitted from medium 2 back into medium 1.

into medium 1 is smaller than the last, and the infinite series of partial waves makes up the resultant reflected wave.

From a macroscopic point of view, the quantities of interest are the amplitude of the incoming wave and the amplitude of the resultant outgoing wave. For the reflectometry technique, we are interested in the intensity, or the square of the amplitude. For the ellipsometry technique, we are interested in the phase and amplitude relationships between the p-wave and the s-wave.

The ratio of the amplitude of the outgoing resultant wave to the amplitude of the incoming wave is defined as the *total reflection coefficient*, and is analogous to the Fresnel reflection coefficients for a single interface. For a single film (two interfaces) this is

$$R^{\mathrm{p}} = \frac{r_{12}^{\mathrm{p}} + r_{23}^{\mathrm{p}} \exp(-\mathrm{j}2\beta)}{1 + r_{12}^{\mathrm{p}} r_{23}^{\mathrm{p}} \exp(-\mathrm{j}2\beta)} \qquad R^{\mathrm{s}} = \frac{r_{12}^{\mathrm{s}} + r_{23}^{\mathrm{s}} \exp(-\mathrm{j}2\beta)}{1 + r_{12}^{\mathrm{s}} r_{23}^{\mathrm{s}} \exp(-\mathrm{j}2\beta)} \tag{2.18}$$

where

$$\beta = 2\pi \left(\frac{\mathrm{d}}{\lambda} \right) \tilde{N}_2 \cos \phi_2 \tag{2.19}$$

These equations are derived in Appendix C.

When $k \neq 0$ the Fresnel coefficients, $\tilde{N}_2$, and $\cos \phi_2$ (and hence β) are complex numbers. When $k = 0$, these numbers are real. In general, except for very special circumstances, R^{p} and R^{s} are complex numbers. β is the phase change in the wave, as it moves from the top of the film to the bottom of the film. Hence, 2β is the phase difference between the part of the wave reflecting from the top surface and the part of the wave which has traversed the film twice (in and out).

For multiple films, the expressions on the right-hand side of Eq. 2.18 can be used in an iterative way[16,17] to determine the total reflection coefficient for the entire stack. For the rapid calculation needed in regression analysis,[18] some form of a matrix method is often used.

2.5 ELLIPSOMETRY AND REFLECTOMETRY DEFINITIONS

2.5.1 Reflectance

The reflectance is defined as the ratio of the intensity of the outgoing wave to the intensity of the incoming wave. The total reflection coefficients R^{p} and R^{s} are defined above as the ratio of the amplitude of the outgoing wave to the amplitude of the incoming wave. Hence, the reflectance is the square of the magnitude of the total reflection coefficient, that is,

$$\Re^{\mathrm{p}} = |R^{\mathrm{p}}|^2 \qquad \text{and} \qquad \Re^{\mathrm{s}} = |R^{\mathrm{s}}|^2 \tag{2.20}$$

For a single interface (no film), the total reflection coefficients reduce to the Fresnel reflection coefficients.

Most reflectance measurements are made at normal or near-normal incidence. Under these conditions, all of the cosine terms are equal to unity, and as suggested in Eq. 2.15, there is no distinction between the p-waves and the s-waves.

2.5.2 Delta and Psi

Figure 2.5 shows the p-waves and s-waves and, in general, they are not necessarily in phase. When each makes a reflection, there is the possibility of a phase shift and the shift is not necessarily the same for the different waves. Let us denote the phase difference between the p-wave and the s-wave before the reflection as δ_1 and the phase difference after the reflection as δ_2. We define the parameter Δ, called *delta* (and often abbreviated as *Del*), as

$$\Delta = \delta_1 - \delta_2 \tag{2.21}$$

Delta, then, is the phase shift which is induced by the reflection, and its value can be from -180° to $+180^\circ$ (or alternatively, from zero to 360°).

In addition to a phase shift, the reflection will also induce an amplitude reduction for both the p-wave and the s-wave, and again, it will not necessarily be the same for the two types of wave. The total reflection coefficient for the p-wave and for the s-wave is defined as the ratio of the outgoing wave amplitude to the incoming amplitude and, in general, this is a complex number. $|R^{\mathrm{p}}|$ and $|R^{\mathrm{s}}|$ are the magnitudes of these amplitude diminutions. We define the quantity Ψ in such a manner that

$$\tan \Psi = \frac{|R^{\mathrm{p}}|}{|R^{\mathrm{s}}|} \tag{2.22}$$

Ψ is the angle whose tangent is the ratio of the magnitudes of the total reflection coefficients, and its value can range from zero to 90°.

2.5.3 The Fundamental Equation of Ellipsometry

Whereas $\tan \Psi$ is defined as the ratio of the magnitudes of the total reflection coefficients, and is hence a real number, we define a complex number ρ (rho) to be the complex ratio of the total reflection coefficients, that is,

$$\rho = \frac{R^{\mathrm{p}}}{R^{\mathrm{s}}} \tag{2.23}$$

The fundamental equation of ellipsometry[19] then is

$$\rho = \tan \Psi e^{\mathrm{j}\Delta} \quad \text{or} \quad \tan \Psi e^{\mathrm{j}\Delta} = \frac{R^{\mathrm{p}}}{R^{\mathrm{s}}} \tag{2.24}$$

The magnitude of ρ is contained in the $\tan \Psi$ part and the phase of ρ is contained in the exponential function. The quantities Ψ and Δ (sometimes only $\cos \Delta$) are measured by ellipsometers. These are properties of our probing light beam. The information about our sample is contained in the total reflection coefficients, and hence in ρ. It should be noted that assuming that our instrument is operating correctly, the quantities Δ and Ψ which are measured are always correct. Whether the quantities such as thickness and the optical constants which we deduce are correct or not depends on whether or not we have assumed the correct model. As an example of this, incorrect values of n and k can be deduced if we assume that our material is a substrate when in fact we have a thin layer of one material on top of a substrate of another material.

This simply makes the point that the quantities which ellipsometers measure are Δ and Ψ. Quantities such as thickness are calculated quantities based on an assumed model.

2.6 REFERENCES

1. F. A. Jenkins and H. E. White, *Fundamentals of Optics*, McGraw-Hill, New York, 1957.
2. F. W. Sears, *Optics*, Addison-Wesley, Reading, 1949.
3. J. D. Jackson, *Classical Electrodynamics*, Wiley, New York, 1962.
4. M. Born and E. Wolf, *Principles of Optics*, 4th edition, Pergamon, New York, 1969.
5. J. Strong, *Concepts of Classical Optics*, Freeman, San Francisco, 1958.
6. O. S. Heavens, *Optical Properties of Thin Solid Films*, Dover, New York, 1965.
7. D. S. Kliger, J. W. Lewis, and C. A. Randell, *Polarized Light in Optics and Spectroscopy*, Academic Press, New York, 1990.
8. M. Born and E. Wolf, *Principles of Optics*, 4th edition, Pergamon, New York, 1969, p. 614.
9. J. Geist, A. Russel Schaefer, J.-F. Song, Y. H. Wang, and E. F. Zalewski, *J. Res. National Institute of Standards and Technology*, **95**, 549 (1990).
10. E. D. Palik (editor), *Handbook of Optical Constants of Solids*, Academic Press, New York, 1985.
11. D. E. Gray (coordinator editor), *American Institute of Physics Handbook*, McGraw Hill, New York, 1972.
12. P. E. Ciddor, *Am. J. Phys.*, **44**, 786 (1976); P. E. Ciddor, *Indian J. Pure Appl. Phy.*, **20**, 397 (1982).
13. H. G. Tompkins, *A User's Guide to Ellipsometry*, Academic Press, New York, 1993, p. 13.
14. D. Brewster, *Treatise on Optics*, Longman, Rees, Orme, Brown, Green and Taylor, London, 1831.
15. W. Spottiswoode, *Polarisation of Light*, Macmillan, London, 1874, p. 11.
16. F. L. McCrackin, Natl. Bur. Stand., Tech. Note 479, (1969).

17. H. G. Tompkins, *A User's Guide to Ellipsometry*, Academic Press, New York, 1993, Appendix A.
18. A general reference for regression analysis is: W. H. Press, B. P. Flannery, S. A. Teukolsky, and W. T. Vetterling, *Numerical Recipes*, Cambridge University Press, Cambridge, 1988.
19. R. M. A. Azzam and N. M. Bashara, *Ellipsometry and Polarized Light,* North Holland, Amsterdam, 1977, p. 287.

CHAPTER 3

Optical Properties of Materials and Layered Structures

3.1 THE PHYSICAL MEANING OF THE OPTICAL CONSTANTS

For single wavelength ellipsometry, the optical constants are reasonably simple: a single number for the index of refraction and a single number for the extinction coefficient. If one only considers a single temperature, that is, room temperature, the word *constant* is appropriate. As indicated in the previous chapter, the value of the index of refraction is an inverse measure of the phase velocity through the material, and the extinction coefficient (related to the absorption coefficient) is a measure of how rapidly the intensity decreases as the light traverses through the material. These properties depend on the wavelength, however, and when a spectroscopic range is considered they are far from constant.

For most dielectric materials, the index of refraction spectra in the visible range are reasonably similar, that is, they differ in magnitude but are similar in shape. In this situation, the extinction coefficient is often negligibly small and is taken to be zero. In 1830, Cauchy[1] proposed an equation to describe the index of refraction as a function of wavelength, and modifications of this equation are still used today. Historically,[2] the *dispersive power* of a material was the difference in the index of refraction for the extreme violet ray and the extreme red ray.

If we extend our interest to include the UV and IR light, and include semiconductors and metals, the spectra of the optical constants become much more complex. It is this rich texture of the optical constants which allows techniques such as spectroscopic ellipsometry and reflectometry to distinguish between materials, and to determine how much each material in a stack contributes to the total thickness. The subject of this chapter is to discuss the optical constant spectra of dielectrics, semiconductors, and metals, and to explain the salient features of the spectra for each type of material. To do this, we first consider the *Lorentz oscillator* material as a prototypical material and show the resulting optical constants. We then extend this concept to three different types of material.

3.2 CONVENTIONS AND MORE BASICS

There are several different conventions for displaying the spectra of optical constants (and ellipsometry and reflectometry spectra). The abscissa can be expressed as wavelength, photon energy (typically in eV), or wavenumbers (number of waves in 1 cm). The last of these is usually reserved for infrared spectroscopy and will not be discussed here. The casual user of ellipsometry or reflectometry with an interest primarily in the thickness of the layers of a stack, will often use wavelength. If the interest lies primarily in the material itself, the focus will often be in the UV part of the spectrum and the choice is then to use the photon energy. Correspondingly, some users will describe the optical properties of a material as suggested in the previous chapter (index of refraction n and extinction coefficient k), whereas other users will use a related quantity the dielectric constant $\tilde{\varepsilon}$.

Although most of this book will use the convention of wavelength and n and k, for this chapter and in a few other places it is convenient to use the photon energy and $\tilde{\varepsilon}$ convention. Therefore, we must introduce these and show the relationships between the various quantities.

Suppose our light beam has speed c, wavelength λ and corresponding frequency ν given by $\nu = c/\lambda$. The energy, E, of a photon in this light beam will be given by

$$E = \mathrm{h}\nu \tag{3.1}$$

where h is Planck's constant. The relationship between the energy, expressed in electron volts, and wavelength, expressed in angstroms, is

$$E \cong \frac{12\,400}{\lambda} \tag{3.2}$$

Like the complex index of refraction $\tilde{N}$, the dielectric constant $\tilde{\varepsilon}$ is a complex quantity which characterizes how a material will respond to excitation by an electromagnetic field at a given frequency.

When an electromagnetic wave with field strength $\bar{E}$ interacts with an atom or molecule which is a part of a material, the local field strength at the atom or molecule is different from the field strength in free space because of interactions with the other atoms or molecules in the material. This difference is a material property, and to understand it we need to introduce the dipole moment, the atomic polarizability, the macroscopic polarization and the displacement. This will lead us to the dielectric constant $\tilde{\varepsilon}$ (actually the dielectric function).

Suppose the acting electric field separates the center of the negative charge from the center of the positive charge by a distance $\bar{r}$ forming a dipole. The dipole moment $\bar{p}$ is given by

$$\bar{p} = e\bar{r} \tag{3.3}$$

where e is the electronic charge. Let us assume that the displacement is sufficiently small that a linear relationship exists between the dipole moment and the field. We define the frequency-dependent atomic polarizability $\tilde{\alpha}(\omega)$ to be the proportionality factor such that

$$\bar{p} = \tilde{\alpha}(\omega)\bar{E} \tag{3.4}$$

If we have N oscillators (this N is not to be confused with the complex index of refraction $\tilde{N}$) per unit volume, the macroscopic polarization $\bar{P}$ is given by

$$\bar{P} = N\bar{p} = N\tilde{\alpha}\bar{E} \tag{3.5}$$

The microscopic field at the location of the atom or molecule is called the displacement field $\bar{D}$ and is given by

$$\bar{D} = \bar{E} + 4\pi\bar{P} \tag{3.6}$$

We see that the displacement field is made up of the external field plus the additional contribution induced by the interaction of the external field with the surrounding material.

Since $\bar{P}$ is also a function of the electric field $\bar{E}$, we can write $\bar{D}$ as

$$\bar{D} = \bar{E} + 4\pi N\tilde{\alpha}\bar{E} = (1 + 4\pi N\tilde{\alpha})\bar{E} \tag{3.7}$$

and we define the quantity inside the parentheses as the dielectric function $\tilde{\varepsilon}$ and write

$$\bar{D} = \tilde{\varepsilon}\bar{E} \tag{3.8}$$

The dielectric function $\tilde{\varepsilon}$ represents the degree to which the material may be polarized by an applied external electric field. The dielectric function $\tilde{\varepsilon}$ is related to the complex index of refraction $\tilde{N}$ by the relationship

$$\tilde{\varepsilon} = \tilde{N}^2 \tag{3.9}$$

and as a complex number, is often expressed as

$$\tilde{\varepsilon} = \varepsilon_1 + j\varepsilon_2 \tag{3.10}$$

where ε_1 and ε_2 are the real and imaginary parts. Note that ε_2 equals the power absorbed per unit volume and can be calculated from the band structure, if known.

3.3 THE LORENTZ OSCILLATOR[3]

The Lorentz model for an oscillator is equivalent to the classical mass on a spring which has damping and an external driving force. This classical model is shown schematically in Figure 3.1.

The motion of an electron bound to a nucleus driven by an oscillating electric field $\bar{E}$ is then given by the equation

$$m\,\frac{\mathrm{d}^2\bar{r}}{\mathrm{d}t^2} + m\Gamma\,\frac{\mathrm{d}\bar{r}}{\mathrm{d}t} + m\omega_0^2\bar{r} = -e\bar{E} \tag{3.11}$$

where m is the electronic mass and e is the magnitude of the electronic charge. The first term is simply the mass times the acceleration. The third term, $m\omega_0^2\bar{r}$, is a Hooke's law restoring force, where ω_0 is the resonant frequency (the vibration frequency of the system if there was an initial displacement but no damping or driving force). The second term represents viscous damping and provides for an energy loss mechanism resulting from various scattering mechanisms for a solid.[3,4] The term to the right of the equals sign is the driving force.

We take the field $\bar{E}$ to vary in time as $e^{-\mathrm{j}\omega t}$ and assume that $\bar{r}$ will have the same time variation. The solution to Eq. 3.11 is given by

$$\bar{r} = \frac{e\bar{E}/m}{(\omega_0^2 - \omega^2) - \mathrm{j}\Gamma\omega} \tag{3.12}$$

From Eqs. 3.3 and 3.4, the atomic polarizability $\tilde{\alpha}(\omega)$ is given by

$$\tilde{\alpha}(\omega) = \frac{e^2}{m}\,\frac{1}{(\omega_0^2 - \omega^2) - \mathrm{j}\Gamma\omega} \tag{3.13}$$

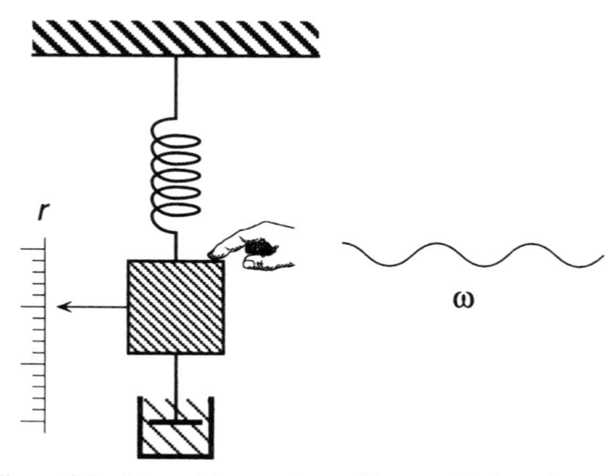

Figure 3.1 Forced harmonic oscillator with damping.

Equations 3.7 and 3.8 define the dielectric function $\tilde{\varepsilon}$ in terms of the atomic polarizability, and hence we obtain

$$\tilde{\varepsilon} = 1 + \frac{4\pi N e^2}{m} \frac{1}{(\omega_0^2 - \omega^2) - j\Gamma\omega} \tag{3.14}$$

Separating the real and imaginary part, we get

$$\varepsilon_1 = 1 + \frac{4\pi N e^2}{m} \frac{(\omega_0^2 - \omega^2)}{(\omega_0^2 - \omega^2)^2 + \Gamma^2\omega^2} \tag{3.15}$$

and

$$\varepsilon_2 = \frac{4\pi N e^2}{m} \frac{\Gamma\omega}{(\omega_0^2 - \omega^2)^2 + \Gamma^2\omega^2} \tag{3.16}$$

If there is more than one oscillator, we can extend Eq. 3.14. Let N_i be the density of oscillators bound with resonance frequency ω_i. Then

$$\tilde{\varepsilon} = 1 + \frac{4\pi e^2}{m} \sum_i \frac{N_i}{(\omega_i^2 - \omega^2) - j\Gamma_i\omega} \tag{3.17}$$

where

$$\sum_i N_i = N \tag{3.18}$$

The frequency dependence of a single oscillator is illustrated in Figure 3.2. Except for a narrow region near ω_0, ε_1, increases with increasing frequency. This is called the normal dispersion. The region near ω_0 where ε_1 decreases with increasing frequency is called the anomalous dispersion.

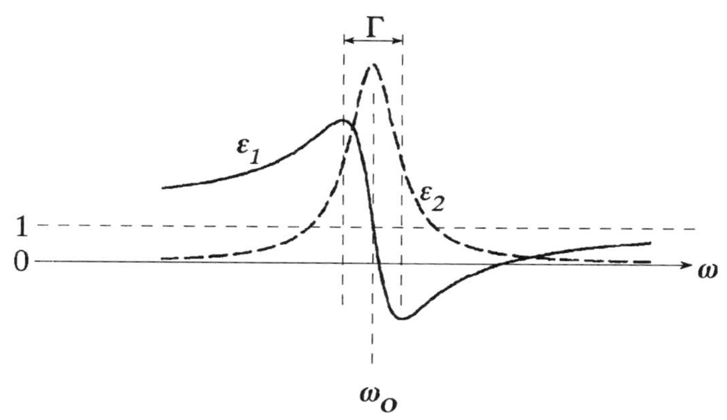

Figure 3.2 The frequency dependence of ε_1 and ε_2 for the Lorentz oscillator.

From the figure and from Eqs. 3.15 and 3.16, we see that ε_2 is always positive and approaches zero for values of frequency far from ω_0, whereas ε_1 may be either positive or negative, and approaches the value of one at frequencies far from ω_0.

From Eq. 3.9, we can obtain values for n and k in terms of the values of ε_1 and ε_2. The relationships are

$$n = \sqrt{\tfrac{1}{2}[(\varepsilon_1^2 + \varepsilon_2^2)^{1/2} + \varepsilon_1]} \tag{3.19}$$

and

$$k = \sqrt{\tfrac{1}{2}[(\varepsilon_1^2 + \varepsilon_2^2)^{1/2} - \varepsilon_1]} \tag{3.20}$$

It must be recognized that the Lorentz oscillator is an idealization. There are several other mathematical descriptions for oscillators which are also frequently used. Real materials correspond to a collection of oscillators grouped together in the UV range, the visible range, and/or the IR range. Nonetheless, the Lorentz oscillator is useful to illustrate the qualitative aspects of insulators, semiconductors, and conductors by observing where the resonant frequency lies in the spectral range.

3.4 INSULATORS

The resonant frequency ω_0 is related to the band gap E_g for the material (i.e., $\hbar\omega_0 \approx E_g$). For materials which we think of as insulators or dielectrics (SiO_2, KCl, etc.), the value of the resonant frequency ω_0 lies well into the UV range (the band gap is high). Figure 3.3, illustrates such a situation, where the resonant energy is 7 eV. If one is observing only the visible radiation, the value of n decreases slightly as the wavelength increases, and the value of k is essentially zero. The values of the index of refraction as a function of wavelength for many insulators[5] can be described in the visible region by a Cauchy function[1] which uses only three parameters. It is given by

$$n(\lambda) = n_0 + \frac{n_1}{\lambda^2} + \frac{n_2}{\lambda^4} \tag{3.21}$$

and the parameters n_0, n_1, and n_2 are called the *Cauchy parameters*. When we use some of the UV range as well as the visible range, the Cauchy relationship is often insufficient to describe the index of refraction, and the value of the extinction coefficient may no longer be negligible. Figure 3.4 shows the index of refraction[6] for several typical insulators. Older tools use the Cauchy function almost exclusively.

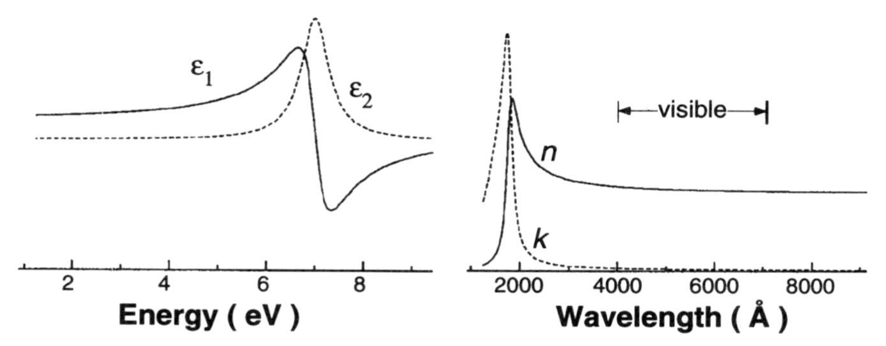

Figure 3.3 Optical functions for a prototypical insulator. The energy corresponding to the resonant frequency of 7 eV. ε_1 and ε_2 are plotted as a function of photon energy, along with the corresponding plots of n and k as a function of wavelength.

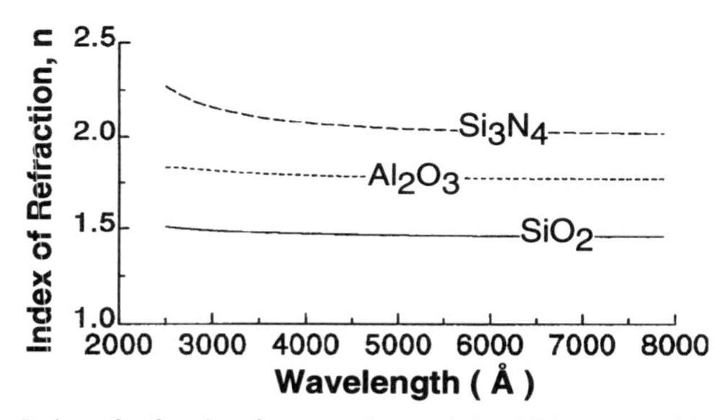

Figure 3.4 Index of refraction for several materials which are considered to be insulators or dielectrics. The data for these curves were taken from Palik.[6]

3.5 SEMICONDUCTORS

Few if any semiconductors can be modeled adequately as a single oscillator. Most require three or more. The basic concepts, however, can be illustrated using a single Lorentz oscillator. The resonant frequency is close to, or in, the visible range (the band gap is less than that of an insulator). The extinction coefficient is small, but nonzero. This is why a silicon wafer is opaque whereas a 500 Å polycrystalline silicon layer on glass will transmit some light. The concept is shown in Figure 3.5 with a prototypical Lorentz oscillator with resonant energy at 3.4 eV.

When using the visible range only, some semiconductors can be modeled with a Cauchy relationship for the index of refraction in much the same way as

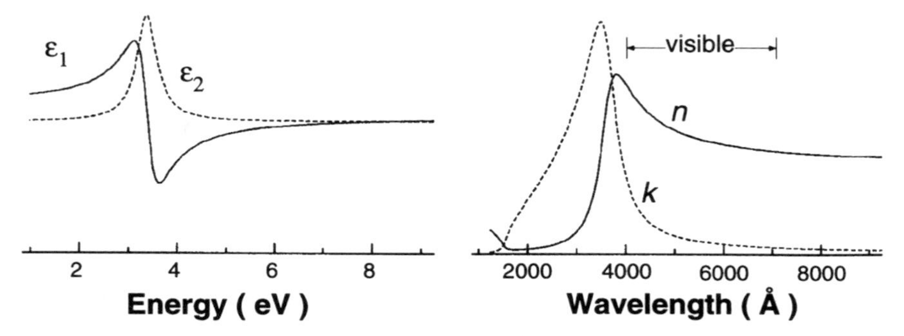

Figure 3.5 Optical functions for a prototypical semiconductor. The energy corresponds to the resonant frequency is 3.4 eV. ε_1 and ε_2 are plotted as a function of photon energy along with the corresponding plots of n and k as a function of wavelength.

for an insulator (see Eq. 3.21). The extinction coefficient is nonzero, however, and must be modeled in some manner or other. In some instances k is modeled with an equation of the same form, that is,

$$k(\lambda) = k_0 + \frac{k_1}{\lambda^2} + \frac{k_2}{\lambda^4} \tag{3.22}$$

where k_0, k_1, and k_2 are called the *Cauchy extinction coefficients*. Some practitioners use exponents of 1 and 3 rather than 2 and 4 in Eq. 3.22. Yet others use the Urbach equation[7] to model k. The Urbach equation is given by

$$k(\lambda) = C_1\, e^{C_2(E-E_b)} \tag{3.23}$$

where

$$E = \frac{12\,400}{\lambda} \qquad \text{and} \qquad E_b = \frac{12\,400}{\lambda_0}$$

and where E is given in electron volts and λ is given in angstroms.

Although there appear to be three adjustable parameters (C_1, C_2, and λ_0), C_1 and λ_0 are completely correlated and hence the Urbach equation is a two parameter equation. The value of λ_0 can be set at a convenient value, often the lowest wavelength measured.

Again, when the UV range is included these simple relationships are insufficient to model the complete curve and the model must include several oscillators, either Lorentz or more complicated ones. This model cannot describe direct-gap semiconductors particularly well. Figure 3.6 shows the index of refraction and the extinction coefficient for the semiconductors Si and InP.

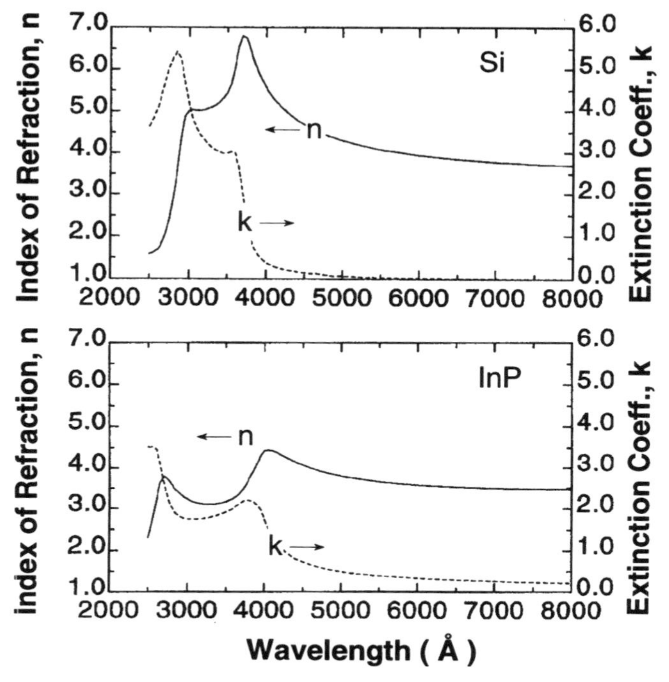

Figure 3.6 The index of refraction and the extinction coefficient of the semiconductors Si and InP. The data for these curves were taken from Palik.[6]

3.6 METALS

Metals can be approximately described by the *Drude model*, which is a modification of the Lorentz oscillator. Since the free electrons in a metal are not bound, we consider the restoring force to be zero. Mathematically, this is equivalent to

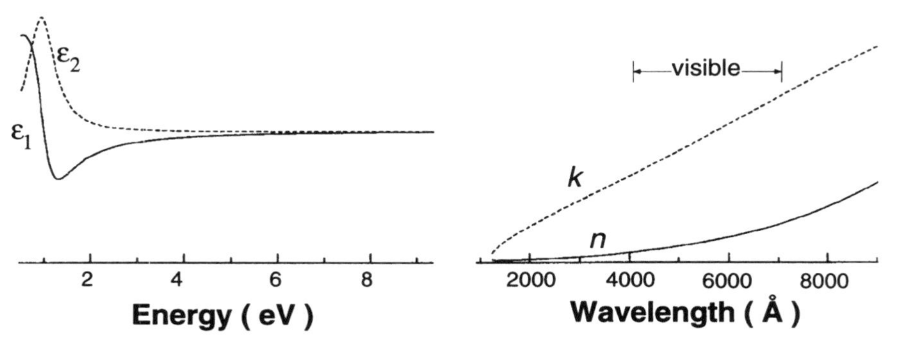

Figure 3.7 Optical functions for a prototypical metal. The energy corresponding to the resonant frequency is near zero. ε_1 and ε_2 are plotted as a function of photon energy along with the corresponding plots of n and k as a function of wavelength.

setting ω_0 to zero. The damping force is still present and represents the ordinary scattering of the electrons associated with electrical resistivity.

In order to illustrate the optical properties of a metal using our Lorentz oscillator model, we simply set ω_0 to a value which is very low (much lower than our frequency range of interest). This is illustrated in Figure 3.7, where we have taken the resonant energy of our prototypical oscillator to be 1 eV. Note that the visible and UV range are now on the other side of the characteristic features of the Lorentz oscillator. Specifically, as the wavelength increases, the values of n and k both increase.

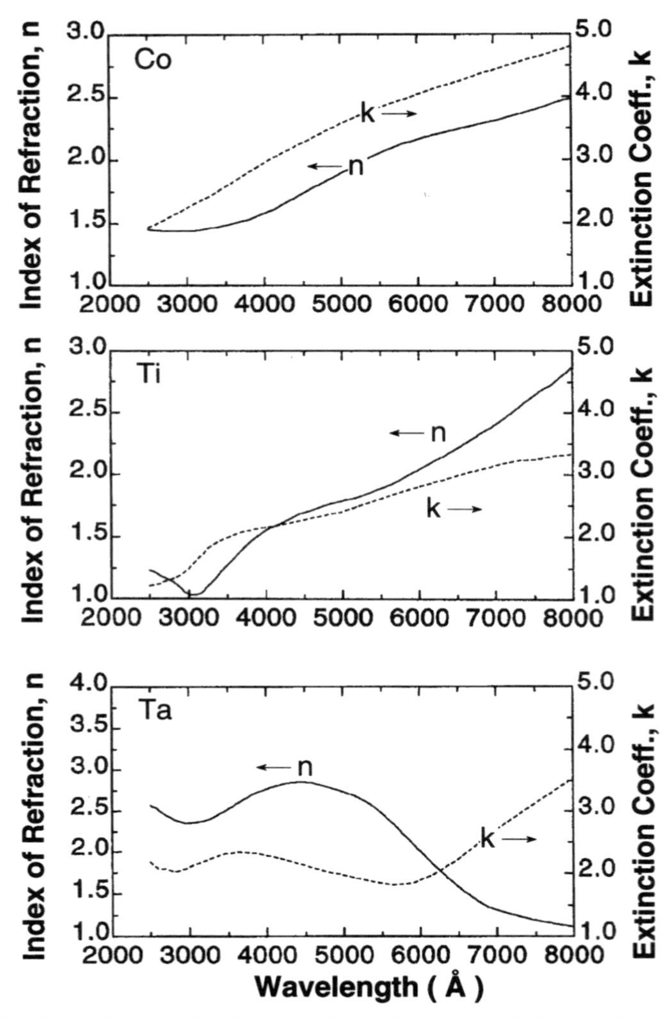

Figure 3.8 The index of refraction and the extinction coefficient of the metals Co and Ti and Ta. The data for these curves were taken from Palik.[6]

Again, no metal is as simple as our prototype. Most have some structure in the visible range. Some have an index of refraction which is less than 1.0 in part of the spectral range. Figure 3.8 shows three metals. The top two are not unlike our prototype, but the lower one is somewhat more complex.

3.7 KRAMERS–KRONIG RELATIONS

In many analysis situations, we use an empirical Cauchy relationship to describe the index of refraction as a function of wavelength. Semiconductors and some dielectrics (e.g., plasma-enhanced chemical vapor deposition silicon nitride) require a nonzero extinction coefficient for a more exact description and it is not unusual simply to use a Cauchy-like relationship to describe the extinction coefficient. Although the Cauchy equations can be rationalized heuristically, we must recognize that this is primarily raw curve-fitting.

From first principles, the index of refraction and the extinction coefficient (or the real and imaginary part of the dielectric function $\tilde{\varepsilon}$) are not independent quantities. The Kramers–Kronig relation relates the two different parts. The relationship between n and k (as a function of energy E) is given by

$$n(E) - 1 = \frac{2}{\pi} P \int_0^\infty \frac{E' k(E')}{E'^2 - E^2} \, \mathrm{d}E' \tag{3.24}$$

or expressed in terms of ε_1 and ε_2,

$$\varepsilon_1(E) - 1 = \frac{2}{\pi} P \int_0^\infty \frac{E' \varepsilon_2(E')}{E'^2 - E^2} \, \mathrm{d}E' \tag{3.25}$$

In other words, if the value of the extinction coefficient (derived from the absorption coefficient) is known over the entire energy range, the index of refraction coefficient can be calculated.

Relationships such as the Lorentz oscillator (Eq. 3.17) satisfy Eq. 3.24 and hence are Kramers–Kronig consistent, whereas the Cauchy relationship (Eq. 3.21 along with either a Cauchy-like equation such as 3.22 or the Urbach equation, 3.23, for the extinction coefficient) does not satisfy Eq. 3.24 and hence is not Kramers–Kronig consistent.

3.8 MISCELLANEOUS

Ellipsometry, and in some instances reflectometry, can be used to determine the optical constants of a material. The quantities determined by these techniques are the thicknesses of layers and the optical constants. In many cases, microstructure information can be determined from the optical constants.

Optical constants are usually presented either as an equation or in tabulated form. Tabulated values for many materials are available in the technical literature and in handbooks. For a material which can be fabricated very reproducibly (e.g., single-crystal silicon or LPCVD silicon nitride), the tabulated values can be used in a straightforward manner, as received. Material properties such as grain size of metal, hydrogen content of PECVD silicon nitride, porosity of deposited oxides, and so on, will affect the optical constants, however, and for these materials, the tabulated values will be similar to the optical constants of the material of interest, but not exactly the same. It is necessary to determine the optical constants of the particular material in hand, along with its thickness.

The treatment of optical constants up to this point in the book has assumed that they are isotropic. For some materials, the index of refraction is different in one direction than another. For a single crystal, this may be determined by the axes of the crystal. For polycrystalline material, it may be determined by preferential orientation of the crystals with regard to the plane of the substrate. Anisotropy plays a very important role in the optical analysis of materials in that many of the optical elements of our measuring instruments are anisotropic. With respect to the material being analyzed, however, we comment that anisotropy adds considerable complexity to the analysis. In many cases, the material can still be analyzed correctly, but the discussion of these methods is beyond the scope of this book. The reader is directed to the instrument software or the technical literature.

We note that in all ellipsometry and reflectometry analysis, the assumption is made that materials are separated by plane parallel interfaces. This is clearly not the case in the real world, where we have intermediate layers between materials and roughness on some surfaces. In the analysis, these are dealt with by assuming that the intermediate layer or the roughness layer is a separate layer which has optical constants intermediate between those of the materials on either side. The method of determining the values of the optical constants of this intermediate layer will be discussed later.

3.9 REFERENCES

1. L. Cauchy, *Bull. des. sc. math.*, **14**, 9 (1830).
2. D. Brewster, *Optics*, Longman, Rees, Orme, Brown and Green, London, 1831.
3. The material in this section can be found in a variety of sources. One is: F. Wooten, *Optical Properties of Solids*, Academic Press, New York, 1972.
4. M. Born and E. Wolf, *Principles of Optics*, 4th edition, Pergamon, New York, 1969, p. 92.
5. W. A. Pliskin, *J. Electrochem. Soc.*, **134**, 2819 (1987).
6. E. D. Palik (editor), *Handbook of Optical Constants of Solids*, Academic Press, New York, 1985; E. D. Palik (editor), *Handbook of Optical Constants of Solids II*, Academic Press, New York, 1991.
7. F. Urbach, *Phys. Rev.*, **92**, 1324 (1953).

CHAPTER 4

Instrumentation

4.1 INTRODUCTION

Reflectometry and ellipsometry are both very old experimental techniques, and there are a great number of different hardware configurations which have been used to perform both experiments. It is impossible to review all of the hardware configurations which have been used to perform reflectance and ellipsometric measurements, so we focus in this chapter on the basic hardware requirements for each type of measurement. Particular emphasis is placed on configurations which have proven commercially successful for research and production applications.

The research and production environments place somewhat different requirements on optical measurement systems. Systems designed for research applications must exhibit accuracy above all other characteristics, while production systems require precision and speed above all else. These requirements have a very strong influence on the particular hardware implementations used for reflectometers and ellipsometers, as will be seen in the following sections.

4.2 REFLECTOMETRY IN GENERAL

The basic principle behind a reflectometry experiment is very simple — the intensity of a beam of light (usually monochromatic) is measured before and after it reflects from the sample under study. The ratio of the intensity of the reflected beam to the intensity of the incident beam is termed the absolute reflectance of the sample. It is generally very difficult to directly measure the intensity of the light beam before it strikes the sample under test, so that most measurements of reflectance involve the measurement of relative reflectance. In a relative reflectance experiment, the intensity of a light beam reflected from the sample under study is divided by the intensity of the same light beam reflected from a known (standard) sample. The ratio of the reflected intensity from the unknown sample to the reflected intensity from the standard sample is termed the relative reflectance, and represents the reflectance of the unknown sample relative to the standard sample.

The absolute reflectance of the unknown sample can always be calculated from its relative reflectance as long as the absolute reflectance of the standard sample is known. For this reason, it is very common to use a bare crystalline silicon wafer as the standard sample. This is because the optical constants of crystalline silicon are well known, and it is a simple matter to calculate the absolute reflectance of the bare silicon wafer from these optical constants. Also, the native oxide on the standard silicon wafer has very little effect on the absolute reflectance of the standard wafer, and is usually stable at around 15–20 Å thickness.

Older reflectometers would often employ aluminum mirrors as standard samples, usually in the form of thick aluminum films coated on glass substrates. These standards have the advantage of being highly reflective over a broad wavelength range, so that the signal-to-noise ratio of the measurement of the standard sample reflectance is very good. Unfortunately, the optical constants of aluminum films vary considerably with deposition technique and conditions, so that it is very difficult to predict the absolute reflectance of these standards from first principles.

4.3 REFLECTOMETER HARDWARE CONFIGURATIONS

A rudimentary reflectometer would consist of a light source, a reflection, and a detector. For spectroscopic reflectometry, either the light source or the detector would be able to separate the light into its various wavelengths. Most production instruments require that the incident light be focused to a small spot, and that the light beam arrives and leaves the sample at normal incidence. The small spot requires the use of a focusing lens, and normal incidence and reflection requires the use of a beamsplitter. A schematic diagram of a simple reflectometer system is shown in Figure 4.1.

In this system, a light beam from the light source is incident on a beamsplitter. Part of the beam is reflected down through a focusing lens (or focusing objective) and imaged onto the sample through the lens. The beam is reflected back from the sample and again passes through the focusing lens. The part of the beam which is transmitted through the beamsplitter then enters the detector, which measures its intensity. Note that the light source may be broadband or monochromatic. If a broadband light source is used, a monochromator may be placed in front of the detector, or a multichannel spectrophotometric detector may be used.

In production systems it is also useful to obtain an image of the sample surface, particularly when patterned samples are under study. It is very common to use the above design with slight modifications, as shown in Figure 4.2, in order to achieve this.

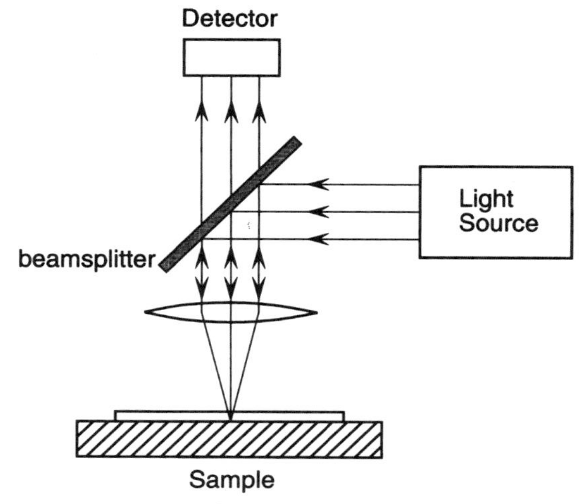

Figure 4.1 A simple single-beam reflectometer.

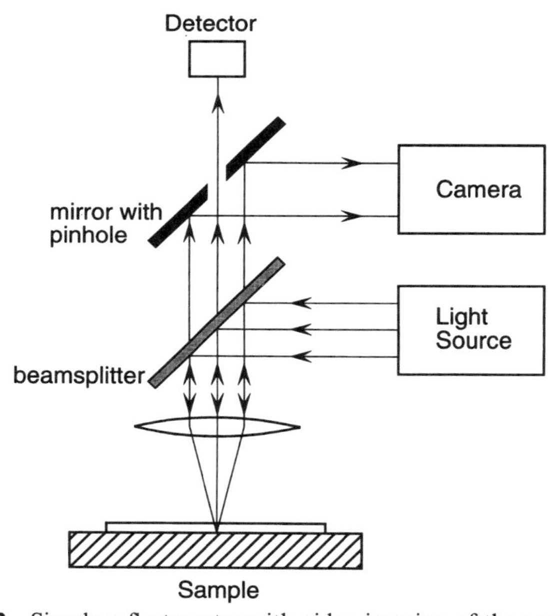

Figure 4.2 Simple reflectometer with video imaging of the sample surface.

In this configuration, a mirror with a small pinhole in its center is placed in the beam path after the beamsplitter. Most of the light beam is reflected from the mirror into a video camera, thus providing an image of the sample surface. A small portion of the beam passes through the pinhole into the detector. This configuration is useful for providing a small spot measurement, and has the added bonus that the portion of the sample from which the reflectance is being measured appears as a small black spot on the video image. This type of system is very common in production reflectometers.

There are two basic types of reflectometers — single beam and dual beam. In a single-beam reflectometer, the reflectance of a standard sample is used to determine the intensity of the incident light beam. In a dual-beam system, the incident light beam is split into two beams, one of which is used to measure the intensity of the incident light. The systems illustrated above are single-beam reflectometers.

A true dual-beam reflectometer would perform a spectroscopic measurement of the intensity of the incident light beam, and would require that the reflectance/transmittance of every component in the light path be known. This is nearly impossible to achieve in practice.

Simpler versions are usually used. One example is simply to add a detector behind the beamsplitter so that the portion of the light beam transmitted through the beamsplitter strikes the added detector, as shown in Figure 4.3 as source monitor (A). This signal is proportional to the total intensity output of the light source, and can be used to correct the measured reflectance for fluctuations in the light source intensity. A variation of this feature is to place an additional detector near the light source in order to take fluctuations in the light intensity into account. This is shown in Figure 4.3 as source monitor (B).

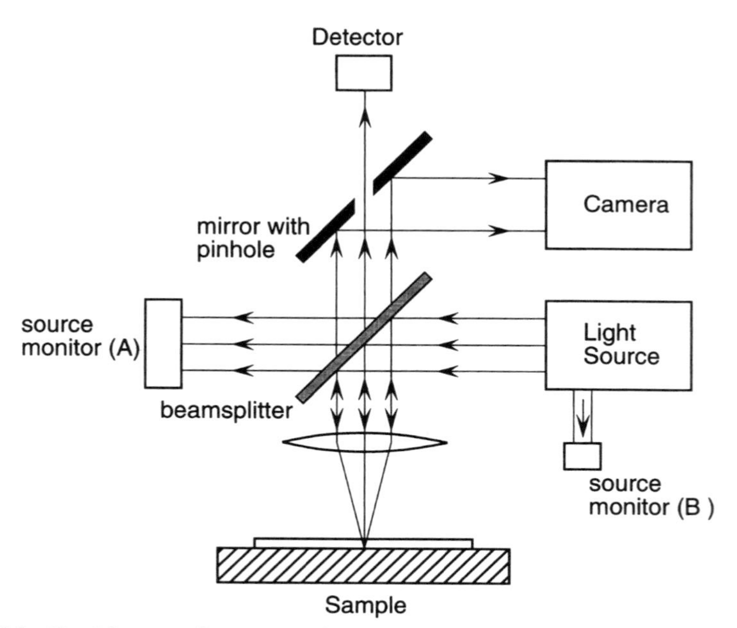

Figure 4.3 Dual beam reflectometer (see text). Two different versions of the source monitor are shown.

This allows a compensation to be made for minor variations in the light intensity. The measurement of the reflectance of a standard sample is still required, however, in order to determine the intensity of the incoming light and hence the absolute reflectance of an unknown sample.

4.4 REFLECTOMETER COMPONENTS

4.4.1 Sources

Almost any type of light source may be used for reflectance measurements, with the primary concern being the stability of the intensity of the output light beam. Arc lamp sources, halogen lamps, and deuterium lamps are probably the most common sources in use. Halogen lamps are often used in tandem with deuterium lamps to provide strong illumination over a broad spectral range (UV–NIR).

4.4.2 Beamsplitter

The choice of a beamsplitter is dictated primarily by the spectral range of the reflectometer. In the configurations shown above, it is desirable for the beamsplitter to split the incident light beam into two beams of equal intensity, as both components of the beam are used for measurements. For visible systems it is possible to use cube or half-silvered mirror beamsplitters. For UV–VIS–NIR systems, beamsplitters consisting of thin film coatings on fused silica or other substrates are often used, as the properties of the beamsplitter may be tuned by optimizing the choice and thickness of the coatings.

4.4.3 Lenses

Most production reflectometers use commercial microscope objectives as the focusing element. This is done in order to provide a high-quality image of the surface of the sample. Also, multiple objectives are usually provided on the system in order to perform measurements of different spot sizes and provide images with a broad range of fields of view.

4.4.4 Monochromators and Detectors

The separation of light into individual wavelengths can be done with a prism or a diffraction grating, or a combination of both. In older systems, a mechanical movement of part of the monochromator was used to allow scanning of the detected wavelength, which was measured by a single detector. Data can be collected much more rapidly if multiple detectors are used to detect all wavelengths simultaneously, after the light has been separated into the various wavelengths. This is the method commonly used at present. A more detailed description of the separation of light into its components is given in a later section in this chapter. Detector systems used for reflectometry are very similar to those used for ellipsometry, which are also described below.

4.5 ELLIPSOMETRY IN GENERAL

Ellipsometry consists of the measurement of the change in polarization state of a beam of light upon reflection from (or transmission through) the sample of interest. Single-wavelength ellipsometry (described in Chapter 6) at a single angle of incidence[1] is used routinely for metrology in microelectronics manufacturing. Ellipsometric measurements may be performed as a function of the wavelength (spectroscopic ellipsometry), as a function of the angle of incidence (multiple angle of incidence, or MAI), or both (variable-angle spectroscopic ellipsometry). In addition, considerable information may be obtained from optically anisotropic samples by acquiring ellipsometric data as a function of the polarization state of the light beam incident on the sample.[2]

A general ellipsometric experiment requires a light source of some sort, something to fix the polarization state of the light beam before it strikes the sample, and something to measure the polarization state after reflection. These simple requirements lead to a few optical components being of primary importance in the construction of an ellipsometer.

4.6 OPTICAL COMPONENTS USED IN ELLIPSOMETERS

4.6.1 Sources

The two most popular light sources for ellipsometric experiments are lasers and arc lamps. Lasers have the advantage of high output intensity and well collimated, Gaussian beams. This is advantageous for systems in which the beam is to be focused to a small spot, and/or in which measurements are performed at or near the Brewster angle for the sample under study. At the Brewster angle, the sample exhibits a minimum in the p-polarized reflectance. In this case the reflected intensity of the light beam may be very small, and the added intensity of the laser source is a definite plus. An example is the measurement of very thin oxide films on silicon at an angle of incidence of 75°. For films which are a few tens of ångstroms thick, the p-polarized reflectance at an angle of incidence of 75° is of the order of 1%, and a rather intense source is required to yield a good signal-to-noise ratio at the detector after reflection of the beam from the sample. A laser will typically have light at one wavelength only; hence arc lamp sources are generally used when spectroscopic measurements are to be performed. These types of sources have the advantage of significant output over a very broad wavelength range. It is much more difficult to obtain a well-collimated light beam from an arc lamp source, however, hence it is much more difficult to focus the beam to a very small spot. Also, arc lamp sources are much less intense than laser sources, making it much more difficult to acquire noise-free data in conditions when the sample reflectance is very low. It should be noted that the regression analysis will be performed across the entire spectrum, however, and therefore small regions of the spectrum with a relatively low signal-to-noise ratio usually do not seriously affect the calculations.

4.6.2 Polarizers

The state of polarization of the incident beam is usually set by a polarizer or a combination of a polarizer and an element called a *compensator* (to be discussed later). Polarizing elements are critical components of any ellipsometric measurement system. As suggested by Figure 4.4, a polarizer is an optical element which converts a light beam of any polarization state (unpolarized, partially polarized, linearly polarized, or elliptically polarized) into a light beam of a single known polarization state. Most polarizers are transmitting in nature, so that the light beam passes through the polarizer, and most polarizers yield a linearly polarized light beam upon transmission through the polarizer.

Linear polarizers are characterized by their optical axis and extinction ratio. When a light beam passes through a linear polarizer, the electric field component of the light beam along the optical axis of the polarizer is unaffected, while the electric field component of the light beam perpendicular to the optical axis is extinguished. The ratio of the parallel to perpendicular electric field components of the transmitted light beam after transmission through the polarizer is defined as the extinction ratio of the polarizer, and is a measure of the quality of the given polarizer. Typical extinction ratios for optical polarizers are 10^6 or greater.

This is to say that regardless of the polarization state of the light when it enters the polarizer, the light which emerges from the polarizer is polarized with the electric field component along the optical axis of the polarizer. If light

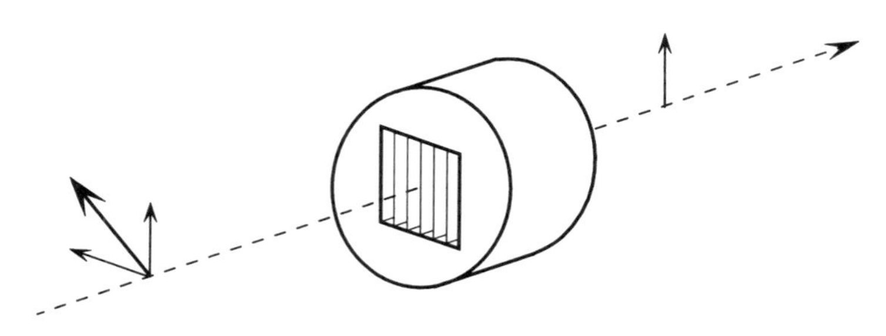

Figure 4.4 An optical polarizer. The polarizer will only pass that component of light which is polarized along its axis.

which is polarized perpendicular to this direction is incident on the polarizer, no light passes through the polarizer (this is referred to as null).

Note that in most ellipsometer systems linear polarizers are used both to polarize the beam incident on the sample and to resolve the polarization state of the light beam after it is reflected from the sample. A polarizer which is used to set the polarization state is usually called a *polarizer* and a polarizer which is used to resolve the polarization state is usually called an *analyzer*. Even though the common name describes the function, they are both polarizers.

4.6.3 Compensators

Many ellipsometer systems also use compensating elements (also referred to as retarders, phase retarders, or sometimes, quarter-wave plates). A compensator is an optical element which alters the phase of one polarization component of a light beam with respect to the other. For example, a 90° compensator (sometimes called a quarter-wave plate) with the azimuthal angle set properly will shift the phase of the s-polarized component of a light beam by 90° while not affecting the phase of the p-polarized component. If a linearly polarized beam of light with equal p- and s-polarized components is incident on a 90° compensator, the resulting transmitted beam will be circularly polarized (illustrated in Figures 2.3 and 2.4). Tompkins[1] gives a detailed discussion of the mechanism whereby a quarter-wave plate achieves this phase shift. Single-wavelength ellipsometers almost always have a compensator. In the early development of spectroscopic ellipsometers most did not have a compensator, although at the time of the writing of this book, compensators are beginning to appear on SE instruments.

4.6.4 Monochromators

The most fundamental device which separates the light into its various components is the prism. As suggested by Figure 4.5, when light moves from air into glass, its direction is changed (the angle of refraction is different from the angle of incidence). The change is dependent on the index of refraction, which in turn depends on the wavelength of the light (if all of the wavelengths had the same index, a prism would not give a spectrum of light). If the two sides of the

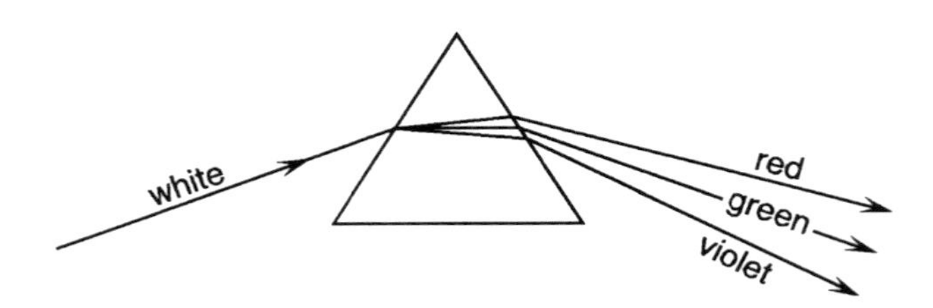

Figure 4.5 Prism separating white light into its components.

glass member are not parallel to each other, the exiting light is still farther separated into its various wavelengths. Sir Isaac Newton (and many people since that time) allowed the light to pass through a prism and fall onto a white surface. The *detector* was the eyeball observing the different colors at the different locations on the white surface.

Another device which will separate white light into its different wavelengths is a diffraction grating, as suggested in Figure 4.6. Although most diffraction gratings in commercial instruments are based on reflection, for simplicity of illustration we use a transmission grating.

A transmission grating consists of a series of parallel slits which are spaced at regular intervals, with spacing d which is comparable to the wavelength of the light λ. A given wavelength will have constructive interference at several given angles of deviation from the incident direction. Equation 4.1 gives the diffraction angle θ for a given wavelength. The integer m has values of 0, 1, 2, ..., and the various beams are referred to as the 1st order, 2nd order, and so on, diffraction beams.

$$\sin\theta = m\,\frac{\lambda}{d} \tag{4.1}$$

Let us consider the 1st order beam only. Since the angle of deviation depends on the wavelength, the light in the 1st order diffraction beam will be separated into the various wavelengths in much the same way as with a prism.

A reflection grating works in essentially the same manner except that the light is reflected from a series of parallel mirrors with the same dimensions and spacing as the slits in the transmission grating. From a structural strength point of view, a reflection grating is much easier to make.

A difficulty arises from the multiple-order diffracted beams. From Eq. 4.1, we see that the light for the 1st order diffraction beam for wavelength λ will occur at the same angle as light from the 2nd order beam for wavelength $\lambda/2$. In order to separate the light from the different orders, a diffraction grating is usually used in conjunction with a filter or a prism.

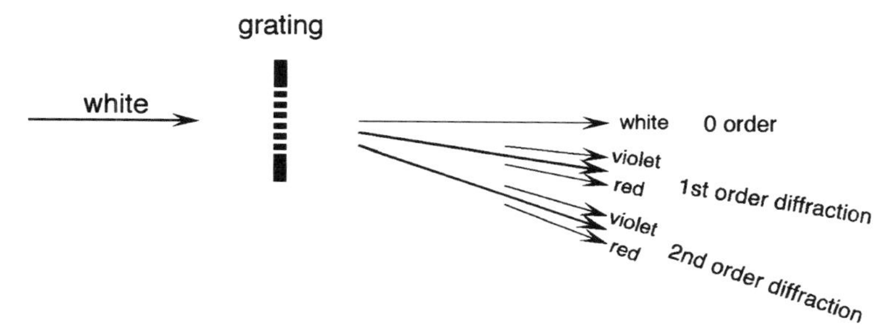

Figure 4.6 Transmission grating separating white light into its components.

4.6.5 Detectors

There are three basic types of optical detectors which are employed for ellipsometric measurements — photomultiplier tubes, semiconductor photodiodes, and CCD arrays (charge-coupled device). Photomultiplier tubes were used for many years as the primary detector. However, they require high-voltage power supplies and exhibit significant nonlinearity. Photomultiplier tubes may also exhibit significant polarization sensitivity. Semiconductor photodiodes have significant advantages in that they are inexpensive, very linear over a broad range of intensity levels, and reasonably sensitive over a broad spectral range. The most commonly used detectors for UV–VIS applications are silicon photodiodes, while InGaAs and HgCdTe detectors are often used for NIR and FTIR applications, respectively. Photodiode detectors may also be fabricated in arrays, which is of particular importance for the construction of production-type ellipsometers where speed of data acquisition is important. This allows multiple wavelengths to be detected simultaneously rather than sequentially. A CCD array is a specific type of semiconductor detector array which is beginning to become more popular for ellipsometric applications. CCD arrays have been used extensively in imaging applications in the past, but have not been as popular for ellipsometry due to problems in correctly interpreting the signal from the detector elements. Advances in CCD design and modeling techniques for data from CCD arrays have led to an increase in popularity of these elements for ellipsometer detectors.

4.7 ELLIPSOMETER HARDWARE CONFIGURATIONS

While the basic ellipsometer configuration is very simple (light source — polarizer — sample — polarizer — detector), there are a large number of ways that an ellipsometric measurement system can be realized. The primary differences between the various ellipsometer configurations occur in the sections which polarize the input beam and resolve the polarization state of the reflected beam. In the next few sections some of the more popular ellipsometer configurations are discussed. Unless noted otherwise, it is assumed for all configurations that a monochromatic light source is used (either a laser or a white light source with a monochromator), and that the measurement is performed at a fixed angle of incidence. Any of these configurations can be implemented in a spectroscopic, variable angle-of-incidence manner.

4.7.1 Null Ellipsometry

One of the oldest ellipsometer configurations is the null ellipsometer. Interestingly, this configuration is also one of the most accurate and robust. The configuration of a null ellipsometer is shown in Figure 4.7.

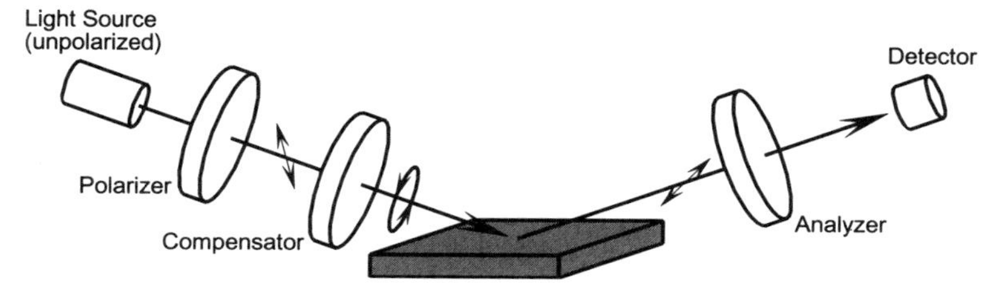

Figure 4.7 Schematic diagram of a nulling ellipsometer.

In a nulling ellipsometer, the light beam passes through a linear polarizer and possibly a compensator before striking the sample under test. After reflection from the sample, the beam then passes through another linear polarizer before striking the detector. In the early years of ellipsometry, nulling ellipsometers were operated manually, with the user adjusting the rotational position of the polarizer and analyzer (another polarizer) in order to completely extinguish (null) the signal from the detector. The compensator was usually held at a fixed angle. The operator then wrote down the azimuthal positions of the three optical elements. Calculations of delta and psi and the subsequent regression analysis were often done with a main-frame computer. Due to the manual nature of these machines, operation of a nulling ellipsometer is very slow. If performed carefully and with four-zone averaging, null ellipsometers can give very accurate experimental data with few or no systematic errors. The operation of the manual null ellipsometer is described in somewhat more detail elsewhere.[1]

With the development of microprocessors, the finding of the null could be automated and the subsequent calculations could be incorporated into the instrument (the rotating element null instrument). This provided much greater speed, and this type instrument has been used extensively for metrology of single layers on silicon or GaAs.

4.7.2 Rotating Element Photometric Ellipsometry

The next evolution of ellipsometry beyond the nulling type systems was the incorporation of continuously rotating elements—usually polarizers or compensators. Whereas for the rotating element null instrument the polarizers were rotated sequentially until null was found, in the photometric instruments, one of the elements rotates continuously during the entire measurement process.

Incorporation of a continuously rotating element into the ellipsometer system also allows automated acquisition of the ellipsometric data, and this type of instrument is somewhat faster than the rotating element null instrument.

There are two common rotating element configurations: rotating polarizer (RPE) and rotating analyzer (RAE). In each type of system, the detector signal is usually measured as a function of time and is then Fourier analyzed in order to obtain the ellipsometric parameters psi and delta.

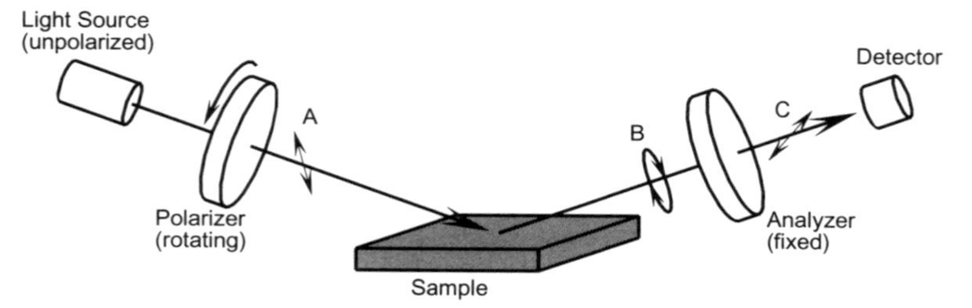

Figure 4.8 Schematic diagram of a rotating polarizer elipsometer.

In the rotating polarizer configuration, shown in Figure 4.8, the input polarizer is continuously rotating while the analyzing polarizer (or analyzer) is held fixed during the measurement. Before it reaches the first polarizer, the light is unpolarized. At location A, the light is linearly polarized and the direction of polarization varies with time (as the polarizer turns). At location B, after the reflection, the light is elliptically polarized and the ellipticity changes as a function of time (as the polarizer turns). After passing through the analyzer at location C, the light is linearly polarized and the direction of polarization is fixed, but owing to the change in the ellipticity at location B, the amplitude at C (and hence the intensity) varies as a function of time (as the polarizer turns). This configuration has the advantage of being unaffected by the polarization sensitivity of the detector, as the polarization state of the beam incident on the detector does not change during a measurement. It also has the disadvantage of being sensitive to any residual polarization of the light beam incident on the initial rotating polarizer. To function accurately, the initial light beam should be completely randomly polarized when incident on the rotating polarizer. This yields a linearly polarized light beam after the rotating polarizer, which exhibits constant intensity as a function of time.

In the rotating analyzer configuration (Figure 4.9), the input polarizer is held fixed while the analyzing polarizer rotates continuously during a measurement. The state of polarization before the reflection at A, is linear, and after the reflection at B it is elliptical, and in both cases, these do not change with time. After passing through the analyzer at C, the light is linearly polarized and the

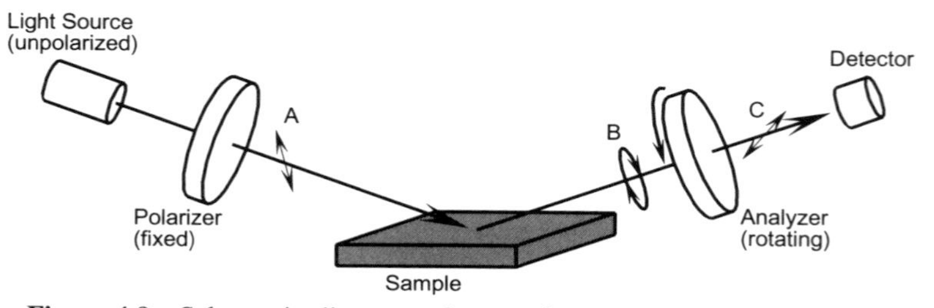

Figure 4.9 Schematic diagram of a rotating analyzer elllipsometer.

direction of polarization varies in time. In addition, the amplitude varies in time (as the analyzer turns) and hence the intensity which is measured by the detector, varies in time.

This type of ellipsometer has the advantage of being insensitive to the polarization state of the light beam incident upon the first (fixed) polarizer, but may suffer from errors due to the polarization sensitivity of the detector. When polarization-sensitive detector systems are used in rotating analyzer ellipsometers, a procedure is usually employed to measure the detector polarization sensitivity and compensate any measured ellipsometric data for this effect. This configuration has been used for many years for single-wavelength ellipsometry[1] for routine metrology. For SWE, a removable quarter-wave plate is also typically used.

One additional feature to be discussed for both a rotating polarizer instrument and a rotating analyzer instrument is the azimuthal angle of the element which does not rotate. For a SWE rotating polarizer instrument, the fixed polarizer is usually set at 45° from the plane of incidence and is not changed. However, the signal-to-noise ratio can be improved if the position of the fixed polarizer is set to a value which is near the value of Ψ. For an SE instrument which measures Δ and Ψ sequentially for many wavelengths, when the monochromator has moved to the next wavelength, the azimuthal angle of the fixed element is typically set at the same value as the measured value of Ψ for the previous wavelength. The typical operation, therefore, is for the rotating element to rotate continuously and for the fixed element to move to the new location and stop before the data collection begins for each wavelength value. This is called *polarizer tracking* or *analyzer tracking*.

One might expect that a rather rigorous fabrication procedure would be required for the instrument to keep track of the location of the axes of the rotating elements. This is normally not done with a fabrication procedure, however, but instead is done by an operational procedure which is referred to as *calibration* (an unfortunate choice of terminology). With the use of an appropriate sample (where the value of Δ would be midrange), the axis of both the polarizer and the analyzer can be determined by measuring the signal for various azimuthal positions of the polarizer and analyzer. For a rotating element SWE instrument, this procedure is based on the principle that the only situation where the reflected beam is linearly polarized is when the incident beam is polarized in the plane of incidence or perpendicular to the plane of incidence. For SE, the procedure involves a regression analysis, and will be described in detail later in this chapter.

One drawback to both the rotating polarizer and the rotating analyzer instrument not having a compensator is that the uncertainty of the values of Δ is large when the value of Δ is near zero or 180°. SWE typically deals with this problem by inserting a quarter-wave plate to shift the measured value of Δ by 90°, thus moving it out of the insensitive region. For SE, when the value of Δ varies strongly as a function of wavelength (e.g., for a dielectric film on a substrate), one simply ignores the measurements where Δ is insensitive and uses the remaining values.

Another drawback to these two types of instrument is that the quantity which is measured is actually cos Δ rather than Δ, and hence it is not clear whether Δ lies between 0° and 180° or between 180° and 360° (or alternatively between −180° and 0°).

There are a few instruments where the compensator is the rotating element. Two polarizers are also used, but both polarizers are held fixed while the compensating element is continuously rotated. This type of system does not suffer from the polarization sensitivity of the detector or to residual polarization in the input light beam. The RCE configuration also has the advantage of being able to measure the delta parameter over the full range of 0–360°.

The drawback to rotating compensator ellipsometers is that it is very difficult to obtain achromatic compensator elements which can be used spectroscopically, and the alignment of the compensating element has a strong effect on its performance.

4.7.3 Modulation Ellipsometry

The final class of ellipsometer designs which will be discussed is polarization modulation ellipsometers (PME).[3–6] A schematic diagram of a typical polarization modulation ellipsometer is shown in Figure 4.10.

In a PME type of system the input polarizer is held fixed during a measurement and a polarization modulator is used to induce a small modulation of the polarization state of the incident beam. The ellipticity of the polarization at location B then varies as a function of time. The reflection changes the ellipticity and the altered ellipticity also varies with the same function of time. PME systems do not usually use a rotating element to analyze the polarization state of the reflected beam. Typically, a Wollaston prism is used to separate the p- and s-polarized components of the reflected beam, and the intensity of each component is measured with separate detectors. At location D, both beams will vary as a function of time (at the modulator frequency). These systems have the advantage of being potentially very fast and having no moving parts, but they can suffer from temperature drift due to the high temperature sensitivity of

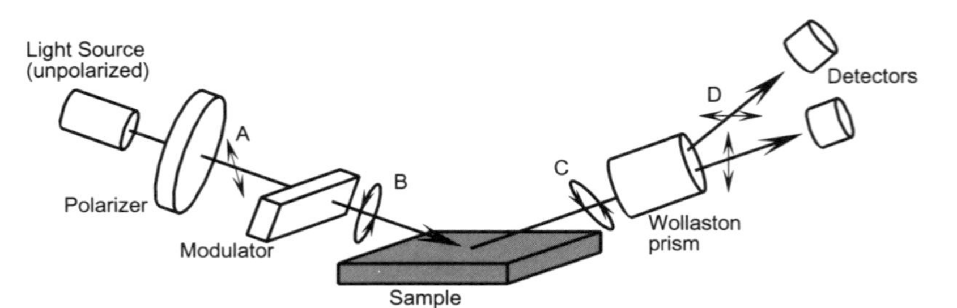

Figure 4.10 Schematic diagram of a polarization modulation ellipsometer.

the polarization modulator. There is no insensitive region for Δ near zero and 180°, and the value of Δ can be determined over the entire range from zero to 360°.

4.8 SINGLE WAVELENGTH ELLIPSOMETERS

Single wavelength ellipsometers (SWE) have been indispensable in the metrology industry for many years.[1] These systems usually employ a laser light source to generate a high-intensity probe beam, with HeNe lasers (6328 Å wavelength) being the most common. Most SWE systems operate at a single fixed angle of incidence, although multiple angles of incidence and variable angles of incidence systems are also available. When the probe beam is focused to a small spot on the sample under test, it becomes possible to perform multiple-angle measurements by profiling the reflected beam. SWE instruments are particularly suited to measuring the thickness and index of refraction of a single transparent layer on a substrate. It is also possible, but more complicated, to measure a single layer in a multilayer stack when the parameters of the other layers are known. The simplicity, speed, and accuracy make this a very important current technology for semiconductor metrology.

4.9 SPECTROSCOPIC ELLIPSOMETERS

Spectroscopic ellipsometers have been a mainstay in the research community for many years, but have only recently become popular in production metrology applications. This is primarily due to the difficulty of focusing a spectroscopic ellipsometer system to a very small spot at the sample surface and to the difficulty of analyzing spectroscopic ellipsometric data. Recent advances in the technology of both the hardware and the software for SE systems, as well as the advent of very powerful and inexpensive personal computer systems, have led to the production of SE systems capable of meeting the stringent throughput and ease-of-use requirements for production metrology tools.

Most SE systems use an arc lamp light source to provide a beam with a significant energy content over a broad range of wavelengths. Research SE systems typically use a scanning monochromator in order to sequentially direct the light for each individual wavelength onto the detector, while production systems typically use a fixed dispersive element to direct the various wavelengths simultaneously onto an array of photodetectors.

Monochromator systems have the advantage of controllable bandwidth, but are inherently slow owing to the necessity of scanning the monochromator across the desired wavelength range. Array detection systems have the advantage of speed, but generally exhibit some degree of smearing of the data due to the nonzero bandwidth of the detection system.

4.10 INSTRUMENT PARAMETER CHARACTERIZATION

In order to acquire accurate data with an ellipsometer, it is first necessary to determine a number of parameters of the optical system, such as the location of the axes of any polarizers and compensators in the system, the polarizing effects due to the focusing optics, the polarization sensitivity of the detector, and the response of the gain circuits on the detector or detector elements. The procedure[7] used to determine these parameters is known as a *calibration* procedure. This terminology has caused considerable confusion in production environments, where the term calibration is often used to describe the scaling of measurements produced by a given tool to match some standard sample or samples. The parameter characterization of an ellipsometer does not usually rely on a known standard sample, but is a fundamental operational procedure which must be performed to characterize the optical and electrical components of the instrument.

This procedure is different for each ellipsometer configuration owing to the different optical components used in each. Since RPE and RAE ellipsometers are most common in production environments, we will briefly describe the instrument characterization procedure most commonly used for these instruments. This procedure was originally published by Johs et al.,[7] and is termed by the authors the *regression calibration* procedure.

In this method, measurements of the Fourier coefficients of the detector signal (from each element in a multichannel system) are found as a function of the azimuthal angle of the fixed polarizer in the system. These measured data are then fit in order to determine the azimuthal angles of both polarizers as well as the gain parameters of the detector amplifier circuits. This procedure is somewhat slow, but it can determine the location of both polarizers to within a few thousandths of a degree if the signal-to-noise ratio of the system is reasonably good. In principle, any sample may be used for this calibration procedure. In practice, it is found that samples which yield a delta value close to 90° at all wavelengths yield the smallest error bars on the calibration parameters. For this reason, a thermal oxide film approximately 300 Å thick on a crystalline silicon substrate is commonly used.

This procedure is very important, as the azimuthal locations of both polarizers must be known in order to calculate psi and delta from the Fourier coefficients of the detector signals. Any error in the measurement of the fixed polarizer azimuth will translate directly into an equal error in psi, and an error in the measurement of the rotating polarizer azimuth will translate into an error in delta.

4.11 RESEARCH VS. PRODUCTION ELLIPSOMETRY

The requirements placed on ellipsometry systems for research and production applications are somewhat different, and as a result, the implementation of research and production ellipsometers tend to be different. For a research ellipsometer, the absolute accuracy of the measured Δ and Ψ data is of paramount importance. Factors like acquisition speed, ease of use, and spot size are of much less importance. Production ellipsometers require high speed, small spot size, easy operation, and easy servicing. As a result, the absolute accuracy of production ellipsometers tends to suffer somewhat in favor of these requirements. The bandwidth effect due to the use of array detection systems has already been mentioned as a source of measurement error.

The requirement of measurements from a small area on the sample surface also has a strong effect on the accuracy of the measurement. To achieve a small spot measurement, the probing light beam is generally focused to a small spot on the sample surface, and then recollimated after reflection from the sample. This leads to two potential sources of systematic error in the measured ellipsometric data. First, the beam will no longer have a single well-defined angle of incidence (Figure 4.11).

Different parts of the focused beam strike the sample surface at different angles of incidence. As a result, the reflected beam will not exhibit a single uniform polarization state, but will exhibit different polarization states as a function of its position in the beam. This effect can be compensated for in the regression analysis by including a numerical convolution over a range of angles of incidence in the model calculation. Figure 4.12 shows the ellipsometric psi and delta data calculated for a 5000-Å thick thermal oxide film on silicon assuming a single angle of incidence (75°) and with a 6° range of angles of incidence around 75°.

The second effect is the change in polarization state induced in the beam by the focusing optics. When a polarized ray of light strikes a surface at a nonzero angle of incidence (as would be the case for a curved lens or mirror) there will be a change in the polarization state of the beam upon transmission or reflection from or through the surface. Ellipsometry measures the change in polarization state of the light beam upon reflection from the sample, so if the focusing optics also cause a change in the polarization state of the beam, this change cannot be separated

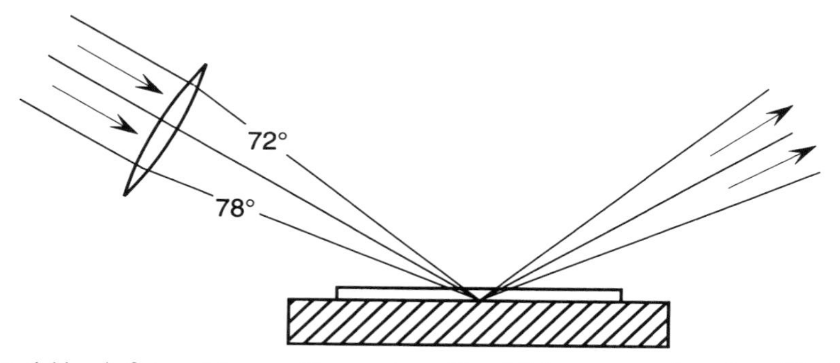

Figure 4.11 A focused beam ellipsometer will exhibit a range of angles of incidence rather than a single well-defined angle of incidence.

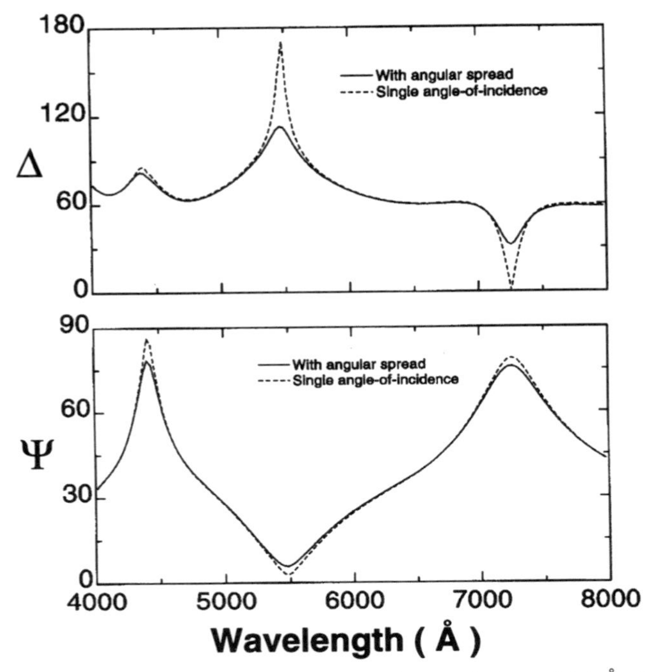

Figure 4.12 Ellipsometric psi and delta data calculated for a 5000-Å thick oxide on silicon. The solid curves were calculated assuming a focused beam with 6° of angular spread, centered around a 75° angle of incidence. The dashed curves were calculated for a single angle of incidence of 75°.

from the change caused by the sample. The change in polarization state due to the focusing optics will depend on the angle of incidence of incoming rays with respect to the surface normal of the optic and on the optical constants of the focusing optic. Glass optics with low numerical aperture show very small polarization effects owing to the low incidence angles of the light rays and the low index of the glass. However, the dispersion of the glass index of refraction causes the focal length of the optic to change with wavelength, thus limiting the spectral range which can be achieved. Reflective optics allow broadband measurements, as the focal length is independent of the wavelength, however, the ray angles are usually significant, and the reflective surface optical constants are generally quite different from air and are strongly wavelength-dependent.

4.12 ERRORS IN REFLECTANCE AND ELLIPSOMETRIC MEASUREMENTS

Errors in both reflectance and ellipsometric measurements may be grouped into two types — random and systematic. Random errors (noise) are typically a function of the signal-to-noise ratio of the system, which is a strong function of the absolute reflectance of the sample under test. Random errors tend to have little effect on the accuracy of the measured data, but may have a very strong effect on the repeatability of the measurements. Averaging is commonly employed to reduce the effects of random errors in both reflectance and ellipsometric systems.

Examples of systematic errors are errors in the calibration of the polarizers, misalignment of the optical elements, errors in the determination of the angle of incidence, and errors in the wavelength calibration of the monochromator or multichannel array. Errors of this type may have serious effects on the accuracy of the measured data, and they cannot be diminished by averaging the measured data. Also, systematic errors will tend not to affect the error bars on the parameters extracted from the measured data, as the statistical analysis of the data is based on the assumption of random errors on the measurement data. For this reason, it is possible to obtain measurements from an ellipsometer or reflectometer which appear to be reasonable (good fit between the model and the measured data, repeatable), but are in fact incorrect. It is therefore very important to characterize the instrument parameters accurately prior to performing measurements in order to minimize systematic errors in the measurements. In production systems (particularly ellipsometers) where it is difficult to eliminate all systematic errors, it is common to simply ignore the systematic errors and adjust the results of the measurements to match some standard sample or samples. While this practice is unacceptable from a research point of view, it can be advantageous to the production user of optical measurement tools, particularly when faced with the task of matching measurement results from a variety of tools.

4.13 REFERENCES

1. H. G. Tompkins, *A User's Guide to Ellipsometry*, Academic Press, New York, 1993.
2. M. Schubert, B. Rheinlander, J. A. Woollam, B. Johs, and C. M. Herzinger, *J. Opt. Soc. Am. A*, **13**, 875 (1996).
3. S. N. Jasperson and S. E. Schnatterly, *Rev. Sci. Inst.*, **40**, 761 (1969).
4. S. N. Jasperson, D. K. Burge, and R. C. O'Handley, *Surf. Sci.*, **37**, 548 (1973).
5. G. E. Jellison, Jr. and F. A. Modine, *Polarization Considerations for Optical Systems II*, 1989, p. 231.
6. G. E. Jellison, Jr. and F. A. Modine, *Appl. Opt.*, **29**, 959 (1990).
7. B. Johs, *Thin Solid Films*, **234**, 395 (1993).

CHAPTER 5

The Anatomy of a Reflectance Spectrum

5.1 GENERAL

In this chapter we discuss the look of a reflectance spectrum and its salient features. We also consider what causes the features to be as they are. The goal of this chapter is for the reader to be able to look at a spectrum and visualize some of the features of the material in the sample. In some cases, the thickness of the film can be extracted directly from the reflectance. In other instances a regression analysis is required and this chapter should provide an understanding of how to choose the seed values for the regression analysis.

Although some UV–VIS spectrophotometers use a nonnormal angle of incidence, the majority of metrology reflectance instruments bring the light to the sample at normal incidence. Accordingly, for this chapter, we shall assume a normal incidence beam. This causes all of the cosine terms in the Fresnel reflection coefficients (Eq. 2.14) to have a value of unity. In addition, the cosine term in the phase equation (Eq. 2-19) is also unity. At normal incidence, there is also no distinction between the s-waves and the p-waves.

The purpose of this chapter is to illustrate general features, and the chapter is not intended to be all inclusive. Accordingly, for the discussion and illustration in this chapter, we will assume that the index of refraction of the film is less than that of the substrate, and that the film is a dielectric, that is, the extinction coefficient is zero. When the structural conditions are different from this, some of the conclusions of this chapter must be modified.

5.2 SPECTRA FOR SUBSTRATES ONLY

When no film is present, the Fresnel reflection coefficient is given by

$$r_{13} = \frac{\tilde{N}_3 - \tilde{N}_1}{\tilde{N}_3 + \tilde{N}_1} \tag{5.1}$$

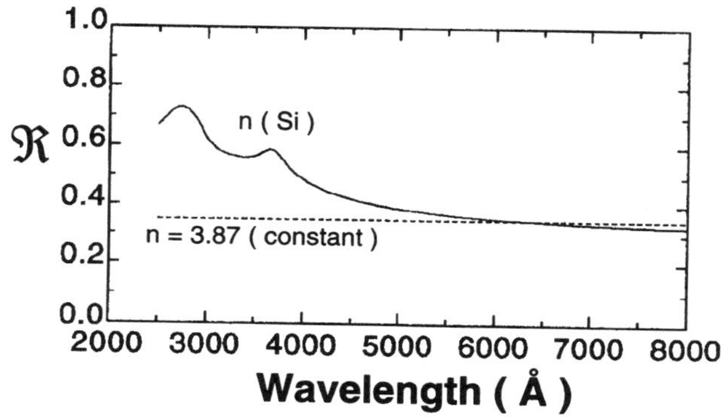

Figure 5.1 The reflectance spectra of a film-free material which has an index of refraction of 3.87 for all wavelengths, along with the reflectance spectrum from a film-free real material (single-crystal Si).

where $\tilde{N}_3$ is the complex index of refraction for the substrate, and $\tilde{N}_1$ and is the complex index of refraction for the ambient and usually equals 1.0. (We reserve the term $\tilde{N}_2$ for the complex index of the film, when we have a film.) The reflectance $\Re$ is given by

$$\Re = |r_{13}|^2 \tag{5.2}$$

To obtain a high reflectance in air, the material must have either an index which is very different from 1.0 or an extinction coefficient which is very different from zero (or both).

For the purposes of illustration, we shall first assume that the index of refraction does not vary with wavelength, and we discuss the effects of wavelength and thickness with this assumption. Then, as a perturbation, we shall consider the effect of optical constants which are functions of wavelength.

If the index of refraction does not vary with wavelength, the reflectance spectrum would be a straight line, calculated from Eq. 5.2. Figure 5.1 shows the reflectance of such a material which has an index of refraction of 3.87 for all wavelengths, along with a typical reflectance spectrum for a real material (where the optical constants vary with wavelength).

5.3 WITH A FILM: THE EFFECT OF WAVELENGTH

We now consider a substrate with a single film. Let us consider how wavelength would affect the values of $\Re$ *if there were no changes in the optical constants as a function of wavelength*. From Chapter 2, we see that at normal incidence, the total reflection coefficient, R, is given by

$$R = \frac{r_{12} + r_{23}\exp(-j2\beta)}{1 + r_{12}r_{23}\exp(-j2\beta)} \tag{5.3}$$

where the Fresnel coefficients are defined analogous to that given by Eq. 5.1. β, the phase change from the top of the film to the bottom of the film, is given by

$$\beta = 2\pi\left(\frac{d}{\lambda}\right)\tilde{N}_2 \tag{5.4}$$

With a film, the reflectance is given by

$$\Re = |R|^2 \tag{5.5}$$

When the film thickness is zero, the exponential terms are equal to +1 and it can be shown that Eq. 5.3 reduces to Eq. 5.1.

For film thicknesses which are small compared with the wavelength, the effect of the exponent (or β) is to cause the reflectance to decrease. This is because the light which reflects from the ambient/film interface, is out of phase with the light which reflects from the film/substrate interface and the intensity of the resultant emerging beam is less than it would have been if the film were absent. This is illustrated in Figure 5.2 for a film and substrate with optical constants which do not vary with wavelength. For this and several subsequent examples, we choose the index of refraction of the film to be $\tilde{N}_f = 1.457 - 0.0j$ and the index of the substrate to be $\tilde{N}_s = 3.87 - 0.0165j$ (these are the values for SiO_2 and Si at a wavelength of 6328 Å). We also show a range of wavelengths from 2500 to 20000 Å for the purposes of illustrating various phenomenon, even though very few instruments have this spectral range.

Note that the amount of reduction of the reflectance depends on wavelength. The value of β in Eq. 5.4 depends on both the thickness d and the wavelength λ. Because the wavelength λ is in the denominator, smaller

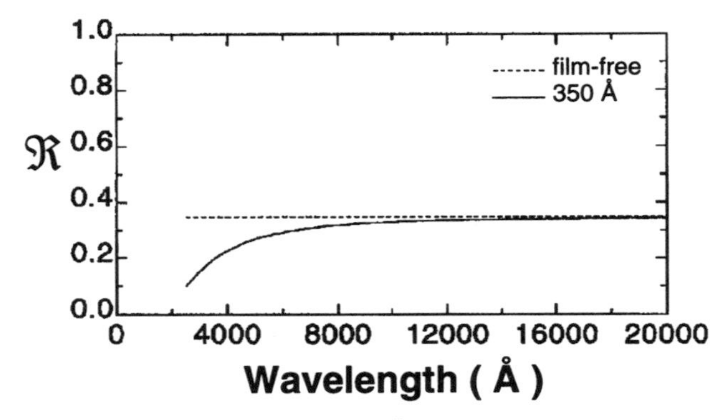

Figure 5.2 The reflectance spectra of a 350-Å film with an index of refraction of 1.457 on a substrate with index of refraction of 3.87 and an extinction coefficient of 0.0165. In this hypothetical structure, the optical constants do not vary with wavelength. The reflectance of a film-free substrate is also shown for comparison.

wavelengths (in the UV) cause greater reductions in the reflectance for a given film thickness. The inclusion of the UV range allowed reflectometers to measure thinner films.

5.4 CONTRIBUTION OF FILM THICKNESS

When the thickness and wavelength are such that the light reflecting from the top surface and the light reflecting from the interface are 180° out of phase, that is, $2\beta = \pi$, the value of the exponent is equal to -1 and we have the greatest amount of destructive interference. This results in the 1st-order minimum, as is shown in Figure 5.3 (for a significantly thicker film than was shown in Figure 5.2). When $2\beta = 2\pi$, the value of the exponent is $+1$, the reflectance is the same as if the substrate were film-free, and we have the 1st-order maximum (shown in Figure 5.3). The minima occur when $2\beta = \pi,\ 3\pi,\ 5\pi,\ 7\pi$, and the maxima occur when $2\beta = 2\pi, 4\pi, 6\pi, 8\pi$.

If one were to have a fixed wavelength and increase the thickness of the film as a function of time, one would pass through the maxima and minima successively. For a fixed film thickness and many wavelengths, we obtain a reflectance spectrum as shown in Figure 5.4 for films which are 3000 and 10 000 Å thick, respectively. Several higher-order maxima and minima are also indicated.

Although we show the wavelength spectral range up to 20 000 Å, many instruments will only measure up to wavelengths of about 8000 Å. On this type of instrument, the 1st-order minimum and the 1st-order maximum would not be observed for the 3000 Å film, and the 1st-, 2nd-, and 3rd-order maximum and minimum would not be observed for the 10 000 Å film.

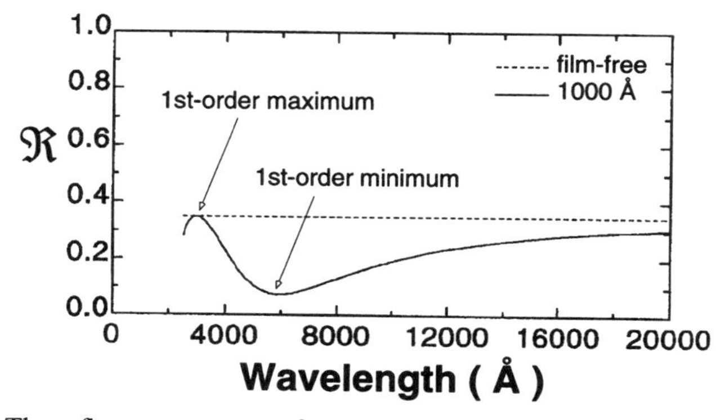

Figure 5.3 The reflectance spectra of a structure like that shown in Figure 5.2 except that the film is 1000-Å thick. The first-order minimum and the first-order maximum are indicated.

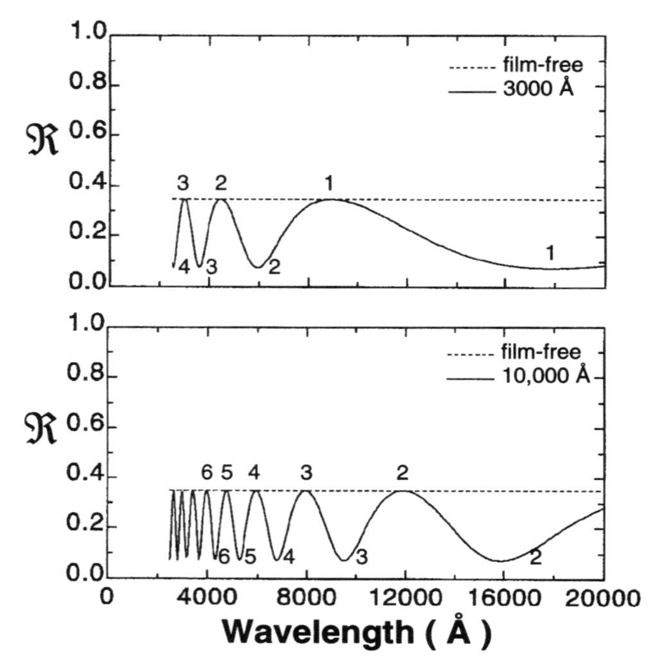

Figure 5.4 The reflectance spectra of structures like that shown in figure 5.2 except that the films are 3000-Å, and 10 000-Å thick, respectively. Several of the higher-order maxima and minima are indicated.

There are several observations which can be made at this point.

1. The maxima (and minima) are further apart for long wavelengths than for short wavelengths.
2. Increasing the thickness causes all of the features to move toward a longer wavelength (to the right in our examples).
3. Thicker films will have more maxima and minima in any given wavelength range than thinner films.
4. For dielectric films, when the index of refraction n_2 is the same for adjacent maxima (as we have assumed for these examples), it can readily be shown that

$$n_2 d = \frac{\lambda_i \lambda_{i+1}}{\lambda_{i+1} - \lambda_i} \tag{5.6}$$

where d is the thickness of the film and λ_{i+1} and λ_i are the wavelengths for adjacent maxima (or minima). When n_2 is a known, slowly varying function of wavelength, this relationship can be used directly to determine an approximate value for the film thickness d. Note that if the photon energy rather than wavelength is used for the horizontal axis, the maxima are equally spaced.

5.5 CONTRIBUTION OF THE OPTICAL CONSTANT SPECTRA OF FILM AND SUBSTRATE

In the above examples, we have made the simplifying assumption that the optical constants did not vary as a function of wavelength. For real materials, the optical constants always vary with wavelength. In some cases, the variation is slight. Examples of this are silicon dioxide and silicon nitride. In other cases, the variation is significant. Examples of this are Si, GaAs, and Cu. To continue our illustration, we add this further complication. Figure 5.5 shows the reflectance spectrum for a 4 μm film of SiO_2 on Si. The curve has many maxima and minima because the film is much thicker than those previously shown. In addition, we show the upper and lower envelope[1] of the curve. In Figures 5.2–5.4, we showed the film-free line which has the appearance of the upper envelope. This is indeed the case. At the maxima, the reflectance is identical to the reflectance for a film-free substrate. Hence for the upper envelope, the reflectance depends on the substrate alone.

When the ambient is air, the reflectance on the upper envelope $\Re_U$ is given by

$$\Re_U = \left| \frac{\tilde{N}_s - 1}{\tilde{N}_s + 1} \right|^2 \tag{5.7}$$

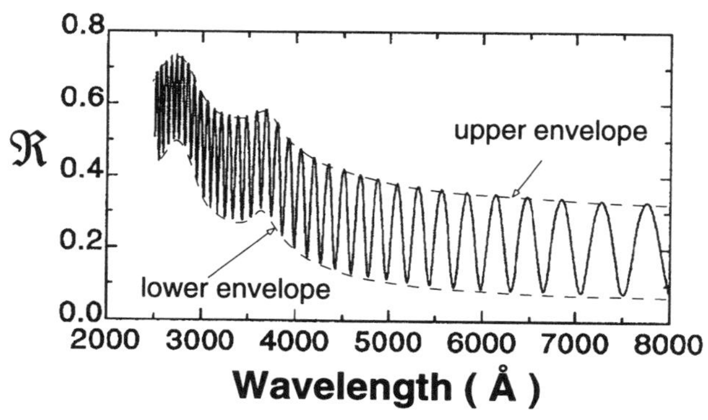

Figure 5.5 The reflectance spectra for a 4-μm SiO_2 film on single-crystal Si. The upper and lower envelopes of the curve are also indicated.

[1] One of the authors (HGT) would like to acknowledge useful discussions with Mark Keefer, KLA-Tencor, about the concepts of envelopes.

where $\tilde{N}_s$ is the complex index of refraction of the substrate (and is a function of wavelength).

At the minima, the value of the exponent in Eq. 5.3 is equal to -1 and we have destructive interference. By algebraic manipulation of Eq. 5.3, it can be shown that the reflectance for the lower envelope, $\Re_L$, is given by

$$\Re_L = \left| \frac{\tilde{N}_s - \tilde{N}_f^2}{\tilde{N}_s + \tilde{N}_f^2} \right|^2 \tag{5.8}$$

where $\tilde{N}_f$ is the complex index for the film (also a function of wavelength). Note that the above considerations do not include the thickness as long as the thickness is such that the reflectance is a minimum.

Figure 5.6 shows the reflectance spectra for films with several different values of the index of refraction on silicon. For films with the same index of refraction as the ambient, no reduction in the reflectance would occur. As the index deviates from that of the ambient, the reduction in the reflectance increases. If the materials are chosen such that

$$\tilde{N}_s = \tilde{N}_f^2 \tag{5.9}$$

at a minimum, the reflectance is zero. The film would then be an antireflective coating. As the index is increased beyond this point, the reduction of the reflectance is decreased until no reduction occurs when the index of the film is the same as the index of the substrate.

Equation 5.8 can be used to estimate the index of refraction prior to determining the thickness. Care must be taken, however, since Eq. 5.8 is a double-valued function. To illustrate this, consider a substrate with index of refraction of 3.87. A film with an index of refraction of 1.967 will be an antireflective

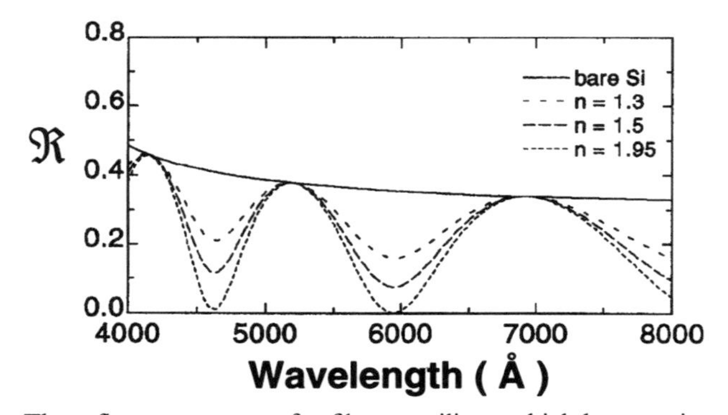

Figure 5.6 The reflectance spectra for films on silicon which have various values of the index of refraction. The thickness of the films were adjusted so that the product of n and d is constant.

coating (zero reflectance) when the thickness is appropriate for a minima. Films with indices of 1.5 and 2.58 will have the same nonzero reflectance values (about 0.07) at a minimum.

5.6 SUMMARY

In this chapter, we have considered the look and feel of a reflectance spectrum for a single layer on a substrate. We have shown that the reflectance of a substrate can be calculated directly from the optical constants of the substrate material. The effect of thin films is to reduce the reflectance, and the amount of reduction is a function of wavelength. For thicker films, there are combinations of film thickness and wavelength which give destructive interference, causing the reflectance to pass through a minimum. We have also seen that there are combinations of film thickness and wavelength which give constructive interference, causing the reflectance to be at a maximum and equal to the reflectance of a film-free substrate. Thicker films give more maxima and minima in a given wavelength range than thinner films, and the maxima and minima are closer together than for thinner films. Adding thickness causes the features to move toward higher wavelengths. Finally, we considered how the location of the maxima (or minima) and the envelopes of the curve can be used to deduce the properties of the film.

CHAPTER 6

Aspects of Single-Wavelength Ellipsometry

6.1 GENERAL

In order to understand spectroscopic ellipsometry (SE), it is necessary to understand some of the concepts of single-wavelength ellipsometry (SWE). SWE has been dealt with extensively in the technical literature and in textbooks[1] and reference books.[2] We will not go into all of the nuances of SWE, but will deal with several aspects which have a profound influence on SE. As mentioned in Chapter 2, the quantities which single-wavelength ellipsometers measure are delta and psi. Quantities such as thickness of films and optical constants of materials are calculated quantities based on a model.

We shall be dealing with substrates and films. A material is defined to be a substrate when we do not have to deal with any lower boundaries or any materials underneath it. A material is considered to be a film if the light which emerges from the top surface is a result of combinations of various reflections from the top and bottom surface of the material. For films, we shall consider both dielectrics ($k = 0$) and absorbing material ($k > 0$). Whereas the primary interest in most ellipsometry is films, it is necessary to have proper optical constants for the substrate in order to obtain information about the film.

An important concept for understanding SWE is the delta/psi trajectory. We shall introduce the concept of the delta/psi domain and then follow the movement of the delta/psi point as the thickness of the film increases. These delta/psi trajectories provide valuable insight about the ellipsometry technique.

6.2 SUBSTRATES: DELTA AND PSI AND THE OPTICAL CONSTANTS

There are several reasons why a given material might be considered a substrate. The most common reason is that the material is absorbing, and the light does not reach the lower interface. A silicon wafer or a thick layer of metal are examples. Another reason might be that light reflected from the lower surface is scattered or diverted so that the reflected light does not go back into the measuring instrument. A glass wafer, where the rear surface is abraded, is an example of such a material.

In SWE, the instrument measures one value for delta and one value for psi. A substrate is the only situation where the numerical parameters can be calculated in a straightforward manner without a regression analysis. Delta and psi can be directly inverted to give one value for n and one value for k. The equation in complex-number format is

$$\tilde{N}_3 = \tilde{N}_1 \tan \phi_1 \sqrt{1 - \frac{4\rho \sin^2 \phi_1}{(\rho + 1)^2}} \tag{6.1}$$

where ρ is the complex quantity including both delta and psi, as described in Chapter 2, and the other quantities are shown in Figure 6.1. If one prefers using real numbers only, this can be separated into

$$n_3^2 - k_3^2 = n_1^2 \sin^2 \phi_1 \left[1 + \frac{\tan^2 \phi_1 (\cos^2 2\Psi - \sin^2 \Delta \sin^2 2\Psi)}{(1 + \sin 2\Psi \cos \Delta)^2} \right] \tag{6.2}$$

and

$$2n_3 k_3 = \frac{n_1^2 \sin^2 \phi_1 \tan^2 \phi_1 \sin 4\Psi \sin \Delta}{(1 + \sin 2\Psi \cos \Delta)^2} \tag{6.3}$$

and these can be further solved for n and k individually. There are multiple solutions for both n and k. The physically correct values will involve positive quantities for both n and k. A key point here is that there is a one-to-one correspondence between the (n, k) values for a material and the film-free (delta, psi) values for that material.

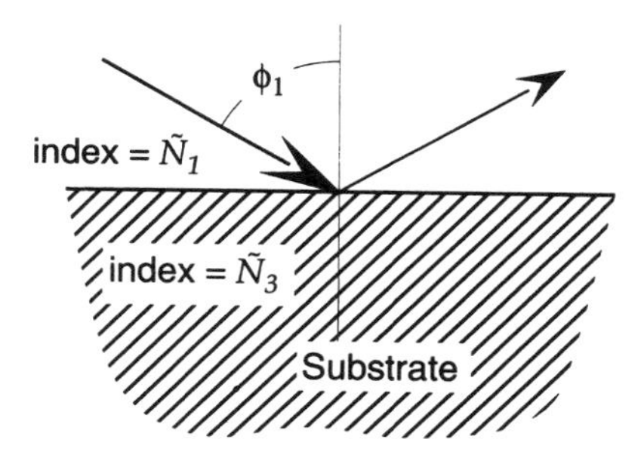

Figure 6.1 Reflection from a substrate.

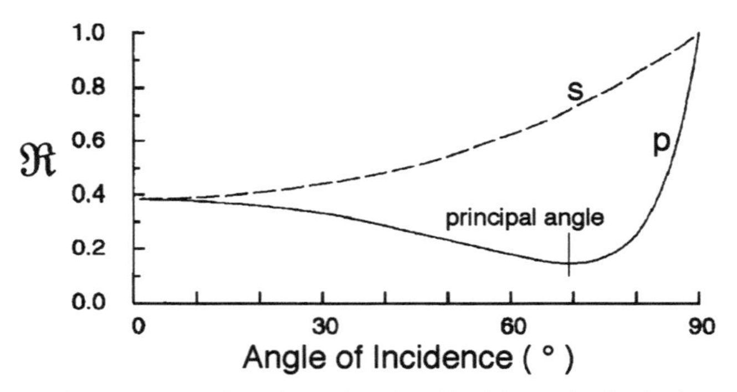

Figure 6.2 Reflectance as a function of angle of incidence for both the s-wave and the p-wave for a material with $k > 0$.

Figure 6.2 (similar to Figure 2.7) shows a plot of the reflectance versus angle of incidence for both the s-wave and the p-wave of a material which has a nonzero value for the extinction coefficient, k. Note that the value for the reflectance of the s-wave is always greater than the value for the reflectance of the p-wave. This implies that for an angle of incidence other than zero or 90°,

$$\frac{|R^{\mathrm{p}}|}{|R^{\mathrm{s}}|} < 1$$

and therefore from Eq. 2.22,

$$\tan \Psi < 1$$

and hence

$$\Psi < 45^{\circ}$$

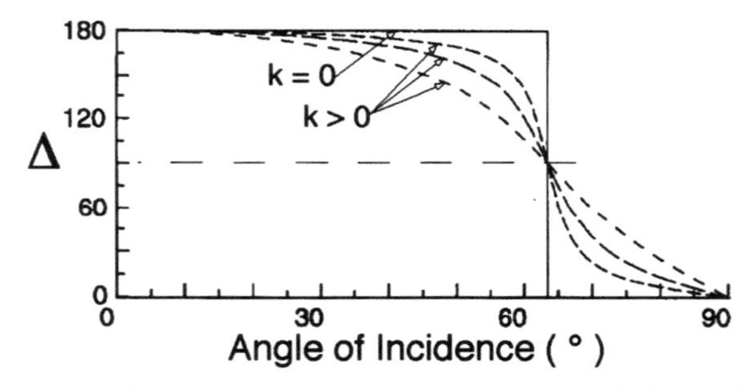

Figure 6.3 Delta as a function of angle of incidence for a dielectric ($k = 0$) and for materials with successively larger value of k. The optical constants are such that the principal angle is the same for all materials.

When light is reflected from a dielectric substrate ($k = 0$), the phase difference between the s-wave and the p-wave is either 180° or zero, depending whether the angle of incidence is above or below the principal angle. As shown in Figure 6.3, as the angle of incidence passes through the principal angle, there is an abrupt shift from 180° to zero.

For a material such as a metal which has $k > 0$, the same transition occurs but is gradual rather than abrupt. The first key point of Figure 6.3 is that for substrates, delta is either 180° or zero for dielectrics and between these two values for metals or any other material with $k > 0$. The second point is that the value of delta passes through 90° at the principal angle. For dielectrics, this transition is abrupt and for other materials it is gradual.

If we pick an angle of incidence of 70° and plot delta against psi, we get the delta/psi domain, as shown in Figure 6.4. In general, for film-covered materials delta can vary from zero to 360° (or alternatively from −180 to +180°), and psi can vary from zero to 90°. From the previous discussion, we see that for film-free material, the delta/psi point will fall in the lower left quadrant only. As indicated earlier, the delta values for dielectrics fall on either the zero or 180° line. In the case shown, all of the dielectrics are

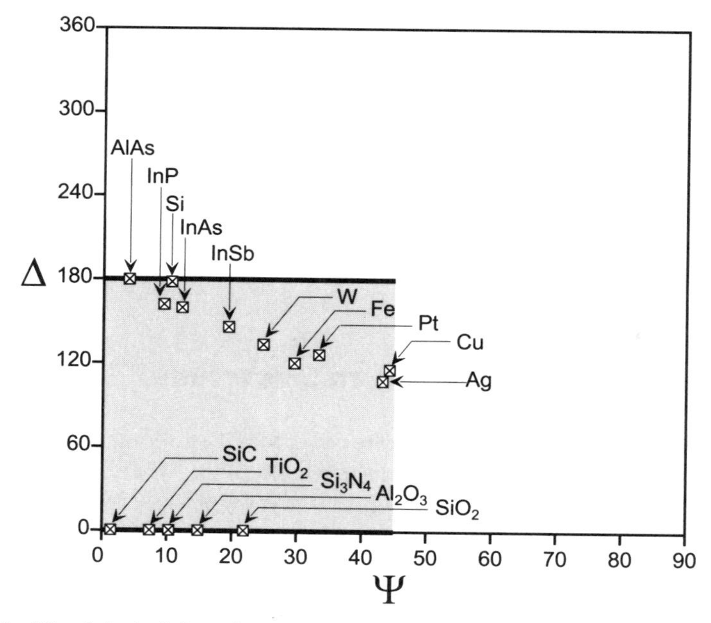

Figure 6.4 The delta/psi domain. The lower left quadrant is where the delta/psi points for a film-free material (a substrate) will be located. The film-free values for several dielectrics, metals, and semiconductors are shown. These values are for an angle of incidence of 70° and a wavelength of 6328 Å.

TABLE 6.1 Optical Constants for Various Materials Along with the Corresponding Film-Free Values of Delta and Psi (for $\lambda = 6328$ Å and Angle of Incidence of 70°)

Material	n	k	Delta	Psi
W	3.64	2.91	134.13	24.62
Fe	2.43	3.30	120.77	29.51
Pt	2.33	4.15	126.99	33.33
Cu	0.25	3.41	107.90	43.14
Ag	0.13	3.99	115.63	44.20
SiC	2.63	0.00	0.00	1.40
TiO_2	2.20	0.00	0.00	7.39
Si_3N_4	2.02	0.00	0.00	10.30
Al_2O_3	1.77	0.00	0.00	14.84
SiO_2	1.46	0.00	0.00	21.67
AlAs	3.11	0.00	180.00	3.95
InP	3.54	0.31	161.95	8.45
Si	3.88	0.02	179.23	10.57
InAs	3.96	0.61	159.24	12.35
InSb	4.28	1.80	146.40	19.19

on the zero line. Had the angle of incidence been 65°, the two left-most points shown would have been on the 180° line. Two of the semiconductors shown, AlAs and Si, have very small extinction coefficients, and hence are very close to the 180° line. For the most part, larger values of k cause the distance from this line to be greater. When delta is near 180°, larger values of n result in larger values for psi.

As indicated earlier, there is a one-to-one correspondence between the film-free delta/psi point of a material and the values of n and k for that material for a given angle of incidence and wavelength. Table 6.1 gives these values for the materials shown in Figure 6.4.

6.3 CALCULATING DELTA/PSI TRAJECTORIES

Let us suppose that we have a substrate covered with a film, as shown in Figure 6.5. We shall distinguish between *transparent* films, where $k = 0$, *absorbing* films, where $k > 0$, but is thin enough that some light reaches the lower interface of the material, and *opaque* films, where the thickness is such that essentially no light reaches the lower interface. Opaque films are usually treated as substrates. With given values of the optical constants of the substrate and the film, one can use the equations in Chapter 2 to calculate the values of delta and psi for any given thickness of the film material. FORTRAN programs are available[1,3] to make this calculation in tabular form, and other software is available[4] to make individual calculations. The calculations could also be

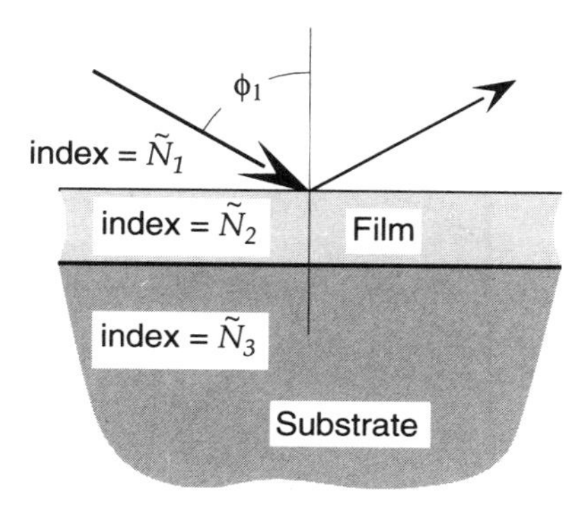

Figure 6.5 A film with index $\tilde{N}_2$ on top of a substrate material with index $\tilde{N}_3$. The ambient has index $\tilde{N}_1$, and is usually air.

programmed into any mathematical software which allows the use of complex numbers. If we calculate the values for a series of thicknesses, the resulting trace in the delta/psi domain is called a *trajectory*. This calculation can be made for single or multiple layers. We shall discuss delta/psi trajectories at length in this chapter, and will find (in later chapters) that there is a similarity between the plot of delta/psi as the thickness is varied for a single wavelength and the plot of delta/psi as a function of wavelength for a single thickness.

6.4 DELTA/PSI TRAJECTORIES FOR TRANSPARENT FILMS

As an example for a transparent film, Figure 6.6 shows the delta/psi trajectory for silicon dioxide on a silicon substrate. The film-free point for silicon is at delta = 178.5 and psi = 10.4, corresponding approximately to $\tilde{N}_3 = 3.86 - 0.04j$. For a film of silicon dioxide, we use $\tilde{N}_2 = 1.46 - 0.0j$. As we begin to add film material, the delta/psi point moves down and slightly to the right. The trajectory is shown as a solid line with values at 100-Å increments shown as circles. An unknown thickness of silicon dioxide on silicon can be determined by simply comparing the measured delta/psi point with the curve shown.

The point crosses the delta = 180° line for a thickness slightly greater than 1400 Å and the value of psi begins to decrease. In this example, the trajectory closes on itself (returns to the film-free point) at a thickness of 2832 Å. For thicknesses greater than this, the trajectory continues down the same path and retraces the entire trajectory. The *film-free point* could also be called the *period point*. This periodicity is one of the drawbacks of single-wavelength ellipsometry, since a given delta/psi point can represent the thickness shown on the curve or it can represent this thickness plus one period, or two periods, or three

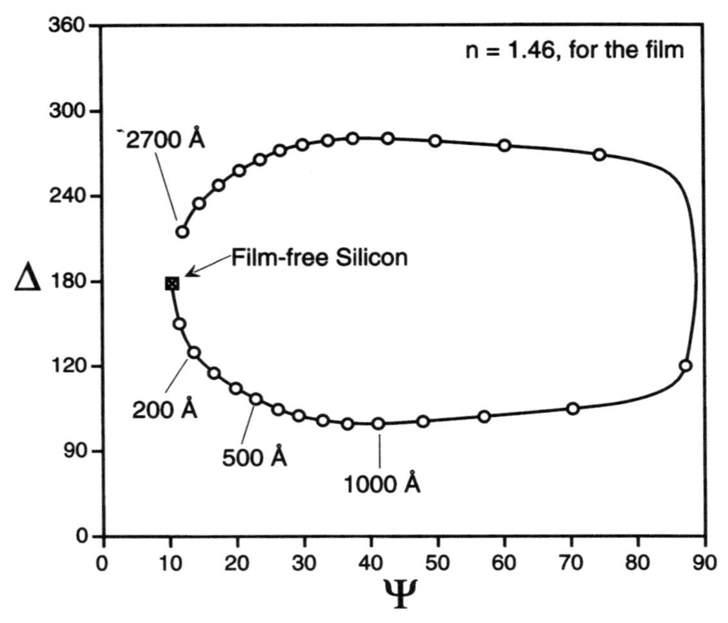

Figure 6.6 The delta/psi trajectory for a transparent film with index $n = 1.46$ (e.g., SiO_2) on a silicon substrate. This trajectory is for an angle of incidence of 70° and a wavelength of 6328 Å.

periods, and so on. The thickness where the trajectory closes on itself (the period thickness) is given by

$$d = \frac{\lambda}{2\sqrt{n_2^2 - n_1^2 \sin^2 \phi_1}} \tag{6.4}$$

which corresponds to $\beta = \pi$ in Eq. 2.19. Note that since the period thickness is a function of wavelength, the difficulty experienced in SWE with the periodicity is a nonissue for spectroscopic ellipsometry and for spectroscopic reflectometry. For a given thickness, the delta/psi point may be close to the period point for a given wavelength. The delta/psi points for other wavelengths, however, will be away from the period point.

The trajectory shown in Figure 6.6 is for one particular index of refraction for the film. Figure 6.7 shows the trajectories up to thickness of about 800 Å for several different indices of refraction. The thicknesses are again shown as circles at 100-Å increments. Note that only one-quarter of the delta/psi domain is shown.

We note that in some regions, the curves are well separated. In this region, *which* curve the delta/psi point falls on determines the index of refraction of the film, and *how far along* that curve the points falls determines the thickness. The

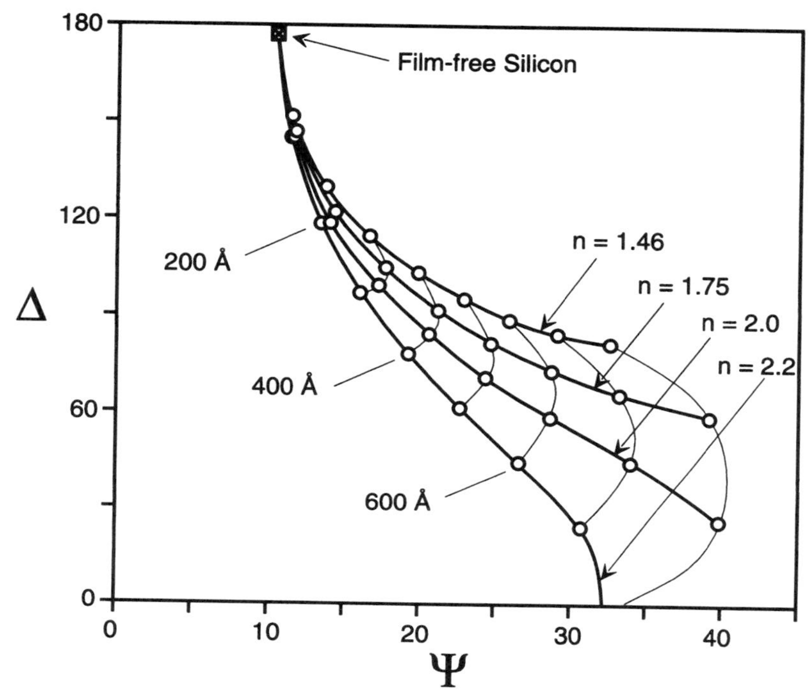

Figure 6.7 The first part of the delta/psi trajectories for transparent films on silicon with the index of refraction of the films as indicated. These trajectories are for an angle of incidence of 70° and a wavelength of 6328 Å.

function of most software in SWE is to make this determination and to convert it into a thickness value and a film index value.

Figure 6.8 shows the complete curves used in Figure 6.7 (without the incremental points). Although several of the curves appear to be discontinuous, one must remember that for any given value of psi, delta $= 0°$ and delta $= 360°$ represent the same point. These same curves are shown on a hemispherical surface in Figure 6.9. Note that the curve for $n = 1.46$ does not go around the pole, and hence passes through the $\Delta = 180°$ line. The other three curves go around the pole and hence pass through the $\Delta = 0°$ line (mostly on the back side of the hemisphere in this figure).

Note in Figures 6.8 and 6.9 that there are regions near the period point where the curves are not well separated. In these regions, small errors in measuring psi will cause large errors in the determination of the index of refraction. In this insensitive region, SWE is not well suited to determining the index of refraction. If one is using the index of refraction as a measure of the quality of the deposited film, it is essential that the thickness be in a region where the measurement is sensitive to index. If the determination of index is not

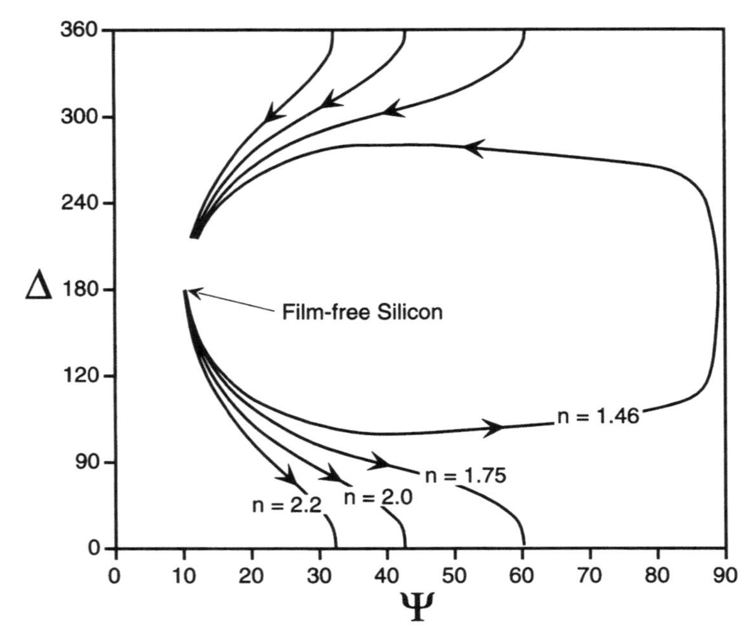

Figure 6.8 The delta/psi trajectories for films with indices of refraction of 1.46, 1.75, 2.0, and 2.2 on single-crystal silicon. Approximately one period is shown.

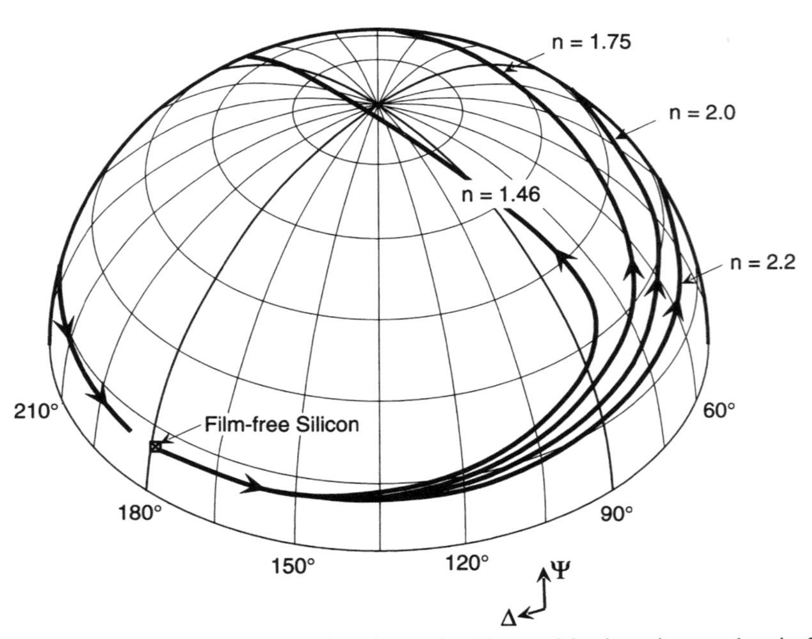

Figure 6.9 The delta/psi trajectories shown in Figure 6.8 plotted on a hemisphere rather than with Cartesian coordinates.

important, and one can assume a value, the thickness can be determined throughout the entire delta/psi domain.

6.5 DELTA/PSI TRAJECTORIES FOR ABSORBING FILMS

Whereas the delta/psi trajectories for transparent films exhibit a periodicity, returning to the film-free values, this is not the case for an absorbing film. For a film-free substrate, the delta/psi point will be characteristic of the substrate material. For a very thick film of an absorbing material, the delta/psi point will be characteristic of a substrate of the film material. The delta/psi trajectory simply moves from one of these points to the other. Figure 6.10 shows the trajectory for a growing film of Ta on a silicon wafer.

We take the film-free point for silicon to be at delta $= 178.5$ and psi $= 10.4$, corresponding to a complex index of $\tilde{N}_3 = 3.86 - 0.036j$. The delta/psi point which corresponds to bulk Ta is at delta $= 94.40°$ and Psi $= 26.96°$, and this corresponds to a complex index[5] of $\tilde{N}_2 = 1.72 - 2.09j$. The film-growth trajectory, therefore, is the movement from the silicon point to the tantalum point. The solid line is the trajectory, with values at 100-Å increments (up to 700 Å)

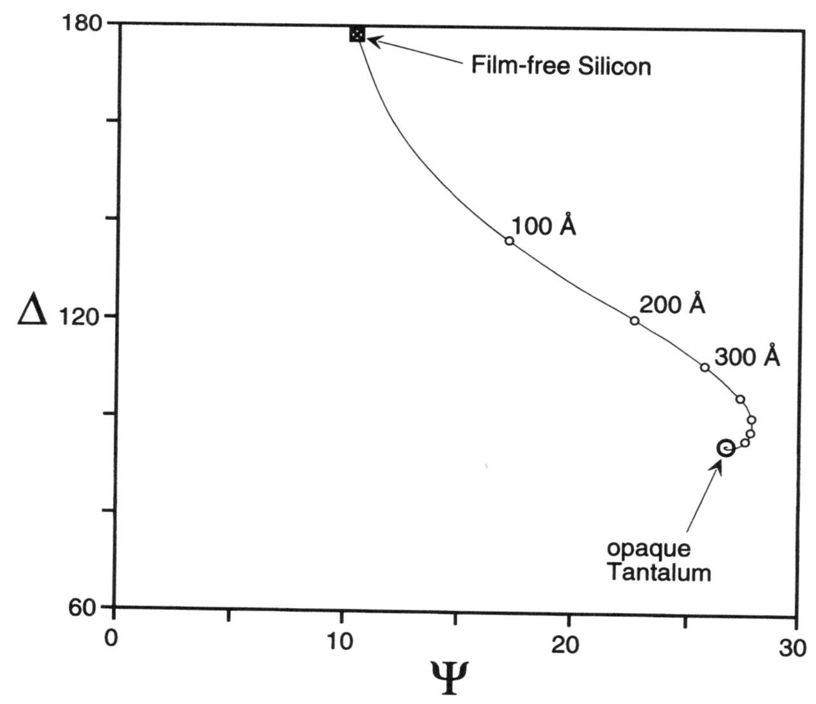

Figure 6.10 The delta/psi trajectory for a film of Ta on a silicon substrate.

being shown as circles. Although not obvious in the figure, the trajectory approaches the opaque tantalum point in an ever tightening spiral (see Tompkins,[1] p. 46).

The delta/psi points are reasonably spread apart for the first few hundred angstroms, and it is possible[6] to use SWE to determine the thickness of metal films in this region. From Eq. 2.9, the thickness where the light intensity has dropped to $1/e$ of its original value (called the penetration depth) is 241 Å. After the thickness reaches about twice this value, the points are so close that it becomes difficult to distinguish one from another. Hence SWE is no longer useful after the film is above about 500 Å. When the film is about four times the penetration depth (in this case, about 1000 Å), the material can be treated as if the film material were infinitely thick (i.e., a substrate).

6.6 TRAJECTORIES FOR TWO-FILM STRUCTURES

Single-wavelength ellipsometry is a technique which is particularly applicable for measuring a single layer on a substrate. However, it is sometimes used for measuring the top layer of a two-layer stack on a substrate. Just as it is essential to know the optical constants of the substrate for a single-layer measurement, it is essential to know the optical constants and the thickness of the underlayer (as well as the optical constants for the substrate) for a two-layer stack. Under these conditions, the measured values of delta and psi can be used to determine the thickness and the index of refraction of the top layer (k is usually assumed to be zero).

The trajectory for a two-layer structure can be illustrated by tracing out the trajectory for the lower layer until the thickness of interest is reached. The trajectory for the upper layer is then started at the current location of the delta/psi point. In Figure 6.11, the dashed line shows the trajectory for silicon dioxide on silicon. Suppose that we stop growing (or depositing) the oxide at a thickness of 1000 Å with the delta/psi point at the location shown by the circle. We then begin depositing a silicon nitride layer, and this trajectory is shown as a solid line. If the growing film were thick enough, the nitride trajectory would close on itself at the starting point of the nitride trajectory, the circle.

Whereas the trajectory for nitride on silicon (shown as $n = 2.0$ in Figure 6.8) crosses the delta $= 0°$ line at about psi $= 43°$, the trajectory for nitride on oxide on silicon crosses this line at about psi $= 85°$. This illustrates that the effect of the underlying film on the nitride trajectory is to offset it.

Note also, that at the transition from oxide to nitride, the direction of the trajectory is relatively unchanged. One of the effects of this is that any uncertainty in the thickness of the underlying layer will be transmitted into uncertainty in the overlayer. This is particularly the case when a nominal value is used and the thickness of the lower layer is not actually measured.

In a manufacturing situation, it is not uncommon to be able to grow oxides to within about ±5%, or in this case 50 Å. If the assumed thickness value was

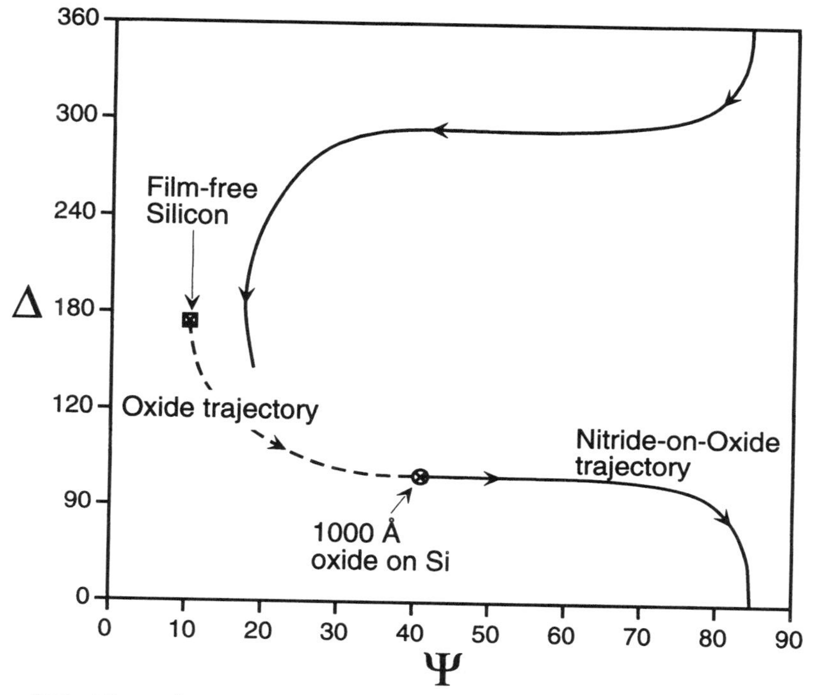

Figure 6.11 The trajectory for a nitride layer on top of a 1000-Å oxide layer on a silicon substrate.

1000 Å and the actual thickness value was 950 Å, the actual nitride trajectory would have started to the left of the assumed location. This "missing" 50 Å of oxide would cause an actual 200 Å nitride to appear to be only 177 Å thick.

When the indices of the two layers are not particularly similar, the trajectory may change abruptly, rather than continue on in roughly the same direction. This is generally the case when the top layer is a metal (see Tompkins p. 49)[1]. When the wrong value for the thickness of the underlayer is used, the delta/psi point does not fall on the calculated trajectory. In this case, the software will often simply give an error message, rather than a thickness value.

6.7 KEY ISSUES

The key issues of this chapter are:

1. For film-free substrates, there is a one-to-one relationship between the delta/psi pair and the n, k pair. For spectroscopic ellipsometry, there is a one-to-one relationship between the delta and psi spectra and the n and k spectra (dispersion relationships).

2. For substrates, delta lies between zero and 180°, inclusive. Psi lies between zero and 45°.
3. For film-covered substrates, as the thickness increases, the delta/psi point traces out a trajectory. For dielectric films, the trajectory closes on itself at the period point and then repeats. For absorbing films, the delta/psi point moves from the substrate material position to the opaque film material position in the delta/psi domain.
4. In general, the location of the trajectory for a dielectric film depends on the index of refraction of the film, and this "spreading" allows the determination of the film index. For very thin films (and for thicknesses near the period thickness) the trajectories are not separated enough to allow an accurate determination of film index.
5. For a two-layer structure, the lower layer causes the trajectory for the upper layer to be offset from its single-layer location. Uncertainty in the thickness of the lower layer is sometimes translated into uncertainty in the determined thickness of the upper layer. In other cases, uncertainty in the lower-layer thickness causes the software to be unable to determine the thickness of the upper layer.

6.8 REFERENCES

1. H. G. Tompkins, *A User's Guide to Ellipsometry*, Academic Press, New York, 1993.
2. R. M. A. Azzam and N. M. Bashara, *Ellipsometry and Polarized Light*, North Holland, Amsterdam, 1977.
3. F. L. McCrackin, Natl. Bur. Stand., Tech. Note 479, 1969.
4. Double Layer Absorbing Film Calculations Program for the IBM Personal System/2, Part Number A9923, Rudolph Research, Flanders, NJ. 07836.
5. E. D. Palik (editor), *Handbook of Optical Constants of Solids II*, Academic Press, New York, 1991.
6. H. G. Tompkins, *Surf. Interface Anal.*, **18**, 93 (1992).

CHAPTER 7

The Anatomy of an Ellipsometric Spectrum

7.1 GENERAL

In this chapter we discuss the look of an ellipsometric spectrum and its salient features. We also discuss what causes the features to be as they are. The goal of this chapter is for the reader to be able to look at a spectrum and to visualize some of the features of the material in the sample. This will allow the analyst to make a better choice of the seed values for the regression analysis.

7.2 TYPICAL SPECTRA

Figure 7.1 shows the spectra of a 4500-Å thermal oxide on silicon taken with an angle of incidence of 75°. We plot the values of Δ and Ψ as a function of wavelength, with Δ ranging from zero to 180° and Ψ ranging from zero to 90°. For single-wavelength ellipsometry (SWE), we would typically show Δ ranging from zero to 360°. Most SWE instruments use a quarter-wave plate and hence can determine the value of Δ in this range.

The quantity which is determined by most spectroscopic ellipsometers is actually $\cos(\Delta)$ and $\tan(\Psi)$ rather than Δ and Ψ, primarily because most spectroscopic ellipsometers do not use a quarter-wave plate. Some analysts choose to display these quantities, and a corresponding plots for the same structures are shown in Figure 7.2.

Whereas Ψ can be determined unambiguously from $\tan(\Psi)$, note that

$$\cos(\Delta) = \cos(360 - \Delta), \quad \text{or } \cos(\Delta) = \cos(-\Delta)$$

so the values of Δ which would normally fall between 180° and 360° are folded into the domain of 0 to 180°.

In Figure 7.1, the value of Ψ undulates between a maximum and minimum, sometimes coming close to either zero or 90°. If the angle of incidence is exactly

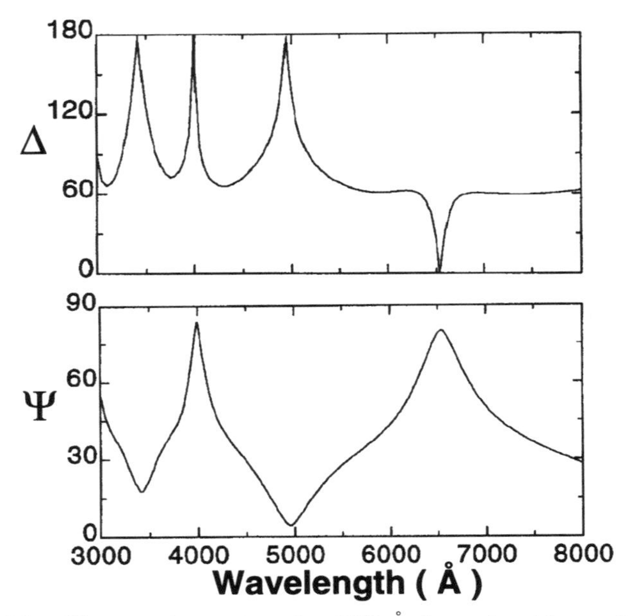

Figure 7.1 The ellipsometric spectra of a 4500-Å thermal oxide on silicon, with an angle of incidence of 75°. Although the quantity which is actually measured is cos(Δ) and tan(Ψ), we plot the vertical axis as Δ and Ψ.

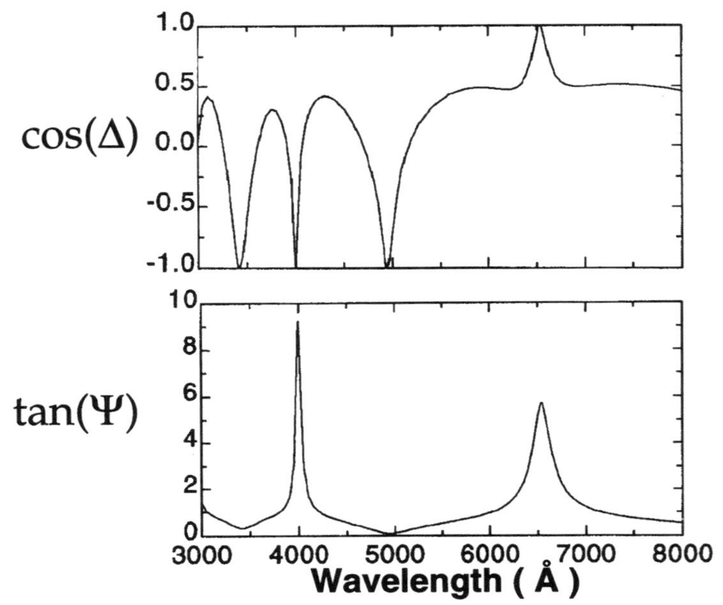

Figure 7.2 The ellipsometric spectra of the same sample as illustrated by Figure 7.1. In this case, we plot the vertical axis as cos(Δ) and tan(Ψ). This convention is used by some practitioners.

at the Brewster angle of the substrate, it is possible for Ψ actually to reach zero, although typically, Ψ never reaches either of the two extremes. As one goes to higher wavelengths, the separation between the extrema increases. For Δ, there are several features in the spectrum which might be described as *cusps*. In some cases, these occur at a value of 180° and in some cases at 0°. We shall discuss these features at length below.

7.3 SPECTRA FOR SUBSTRATES ONLY

When no film is present, there is a one-to-one correspondence between the Δ/Ψ values and the n/k values (or the $\varepsilon_1/\varepsilon_2$ values) of the substrate material. This relationship is given, in complex number format, as

$$\tilde{N}_s = \tilde{N}_a \tan \phi_1 \sqrt{1 - \frac{4\rho \sin^2 \phi_1}{(\rho + 1)^2}} \tag{7.1}$$

where $\tilde{N}_s$ and $\tilde{N}_a$ are the complex indices of refraction for the substrate and ambient, respectively. ϕ_1 is the angle of incidence, and

$$\rho = \tan \Psi e^{j\Delta} \tag{7.2}$$

To determine the optical constant spectrum of a substrate, therefore, we simply measure the values of Δ and Ψ at all the wavelengths being used. The instrument software then converts these values into the appropriate n and k spectra for the substrate. This procedure must be used with care, as surface quality is never perfect and this will affect the results. An unknown surface film or roughness will cause the calculated values of n and k to be slightly different from the true ones.[1]

It was shown in the previous chapter that for a film-free substrate,

$$0 \leq \Psi < 45^\circ \quad \text{and} \quad 0 \leq \Delta \leq 180^\circ$$

A measured value of Ψ greater than 45° implies that the sample is not a simple substrate, but has at least one film on it.

7.4 WITH A FILM: SINGLE-WAVELENGTH CONSIDERATIONS

There is a relationship between wavelength and film thickness which is best understood by considering some aspects of single-wavelength ellipsometry.[2,3] In this section, we show how thickness affects the Δ/Ψ values for a fixed wavelength. In the next section, we will show how wavelength affects the Δ/Ψ values for a fixed thickness.

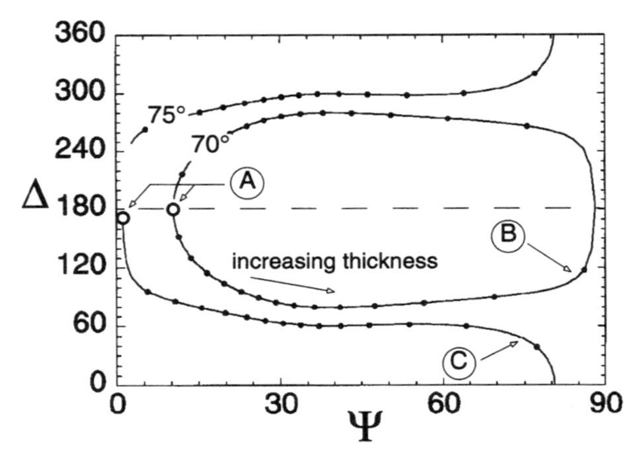

Figure 7.3 The single-wavelength ellipsometry Δ/Ψ trajectory for SiO_2 on Si for angles of incidence of 70° and 75°. The open circles represent the film-free point, and the solid dots are plotted at 100-Å increments.

Figure 7.3 shows the Δ/Ψ trajectory (the plot of Δ vs. Ψ) for SiO_2 on Si as the thickness of the single film increases. The solid points are plotted at 100-Å intervals. In this case, we show the trajectory for a wavelength of 6328 Å at angles of incidence of 70° and 75°. In both cases, the film-free point is near Δ = 180°, and this feature is denoted as A. As the thickness increases from zero, the value of Δ initially decreases and then levels out. For the 70° trajectory, when Ψ becomes larger, Δ begins to increase, passing through 180°. This feature is indicated as B in the figure. For the 75° trajectory, at about the same point in the trajectory, Δ decreases rather than increasing, and this feature is shown as C. Δ passes through zero rather than 180° and re-enters the domain at 360°, eventually moving in a manner similar to the 70° trajectory. Each trajectory then closes on itself (returns to the film-free value) at a thickness called the *period thickness*. In addition to being called the film-free point, this is also called the period point (feature A). The period thickness depends on the wavelength, the angle of incidence, and the index of refraction of the film, but not on the index of the substrate. Note that the trajectory only closes on itself for films with zero extinction coefficient. Δ/Ψ trajectories were discussed briefly in the previous chapter and at length by Tompkins.[2]

The Δ/Ψ domain can be visualized as a hemisphere, with Δ being analogous to longitude (angular distance from some reference) and Ψ being analogous to latitude (angular distance from the "equator"). Two points having Δ of zero and 360° respectively, with the same value of Ψ, represent the same location. A Ψ value of 90° would represent the pole, and all values of Δ on the Ψ = 90° line represent the same point. The observation that the 100-Å increment solid points appear to be farther apart when the trajectory is near Ψ = 90° is analogous to the observation that Greenland and Iceland appear to be large countries on a flat map, and both observations are because of the difficulty of representing a hemisphere on a flat piece of paper.

Utilizing Figure 6.9, let us visualize the two trajectories on a hemisphere. Both start near the longitude of 180°, near the equator. As the thickness increases, both curves move on the hemisphere, with Δ decreasing and moving away from the 180° meridian, and Ψ moving away from the equator and toward the pole (Ψ increasing). The 70° trajectory does not quite reach the pole before it swings back across the 180° meridian (feature B) and begins moving toward the equator, whereas the 75° trajectory passes around the pole. Since it passes around the pole, it must pass through the $\Delta = 0°$ meridian (feature C). Because both trajectories are near the pole, small increments of thickness result in large changes in the value of Δ. This appears abrupt on a flat piece of paper, although it is a smooth transition on the hemispherical representation. This feature of sometimes moving toward zero and sometimes moving toward 180°, depending on the experimental parameters, will be observed in spectroscopic ellipsometry (as a function of wavelength) as well as single-wavelength ellipsometry (as a function of thickness).

7.5 EFFECT OF WAVELENGTH

We now consider how wavelength would affect the values of Δ and Ψ for a constant thickness *if there were no changes in the optical constants.* The total reflection coefficients are given as Eq. 2.18. These consist of the Fresnel reflection coefficients and an exponential term. The Fresnel coefficients are calculated from the optical constants of the various materials. The only place that thickness appears in the total reflection coefficient is in the exponential term. The exponent is $-j2\beta$ where β is given by

$$\beta = 2\pi\left(\frac{d}{\lambda}\right)\tilde{N}_2 \cos \phi_2 \tag{7.3}$$

If, for illustration purposes, we agree that the optical constants are fixed and not functions of the wavelength, then the effect of $1/\lambda$ is exactly the same as the effect of the thickness d. That the optical constants are in fact functions of wavelength will be treated as a perturbation in the following section.

Whereas we often present the result of varying thickness in SWE as a Δ/Ψ trajectory, for SE the result of varying wavelength is usually presented as a Δ spectrum and a Ψ spectrum. For this example, we choose the index of refraction of the film and the substrate to be 1.457 and 3.87, respectively, and the extinction coefficient of the film to be zero and that of the substrate to be 0.0165 (these are the values for SiO_2 and Si at a wavelength of 6328 Å). We choose the thickness of the film as 4500 Å. The resulting spectra are shown in Figure 7.4.

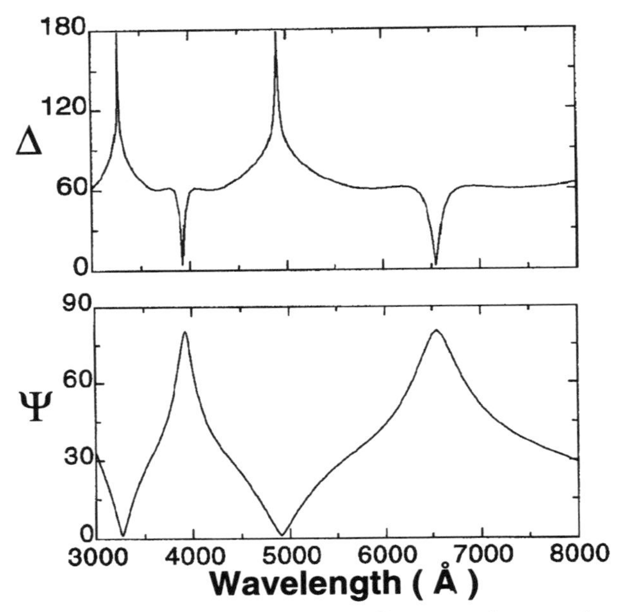

Figure 7.4 The ellipsometric spectra for a single film on a substrate where the values of n and k for both the film and the substrate are constant (do not vary with wavelength). We take $\tilde{N}_f = 1.457 - 0.0j$ for the film and $\tilde{N}_s = 3.87 - 0.0165j$ for the substrate (these are the values for SiO_2 and Si at a wavelength of 6328 Å). The thickness of the film is 4500 Å.

When the value of Ψ is maximum, the value of Δ moves abruptly toward zero. This is analogous to feature C in Figure 7.3, where the trajectory moves to zero and then re-enters the domain at the 360° line. In Figure 7.4, the 180°–360° part has been folded into the 0°–180° domain, and hence the curve appears to have a sharp cusp at this point. For different values of the index of refraction (e.g., > 4.7 for the substrate), these cusps would have been at the 180° line rather than the zero line, corresponding to feature B in Figure 7.3.

In Figure 7.4, when Ψ is minimum, the Δ spectrum goes abruptly to 180° and returns, giving the appearance of another sharp cusp. This corresponds to feature A in Figure 7.3, where the trajectory goes through the 180° line near the period point.

The wavelength separation between cusps of the Δ spectrum (and the maximum/minimum points in the Ψ spectrum) increases as a function of wavelength. The photon energy is proportional to the inverse of the wavelength, and if one were to use a spectral scale of photon energy rather than wavelength, the cusps would be equally spaced in Figure 7.4. The locations where Ψ is minimum corresponds to where β is an integral number of 2π (in Eq. 7.3), and the value of the complex exponential is real and equal to $+1$. The locations where Ψ is maximum correspond to where the exponential is real and equal to -1.

7.6 CONTRIBUTION OF THE OPTICAL CONSTANT SPECTRA OF FILM AND SUBSTRATE

The effect of optical constants which are functions of wavelength is to modify the spectra shown above. Whereas for Figure 7.4 we used the simplification of materials having optical constants which do not vary with wavelength, Figure 7.5 shows the ellipsometric spectra for a 4500-Å SiO_2 layer on Si without this simplification. In this figure, we use the optical constants[4] for SiO_2 and Si (shown in Figure 3.6).

For Ψ, the minimum and maximum spectral locations are about the same as before, but the extreme values of Ψ are modified to some extent. It should be noted that although the spectral location of the Ψ minimum depends on the thickness and optical constants of the film as well as the substrate, the *value* of Ψ at the minimum is dependent only on the optical constants of the substrate and not on the thickness or optical constants of the film. The value of Ψ at the maximum is dependent on the optical constants of the substrate, the film, and the angle of incidence. This property is also evident for a reflectance spectrum and is simpler to put to practical use there, since the angle of incidence for reflectometry is usually zero.

For Δ, the cusp at a wavelength of about 6500 Å is similar to the one in Figure 7.4. The cusp at a wavelength of about 4000 Å now turns toward the

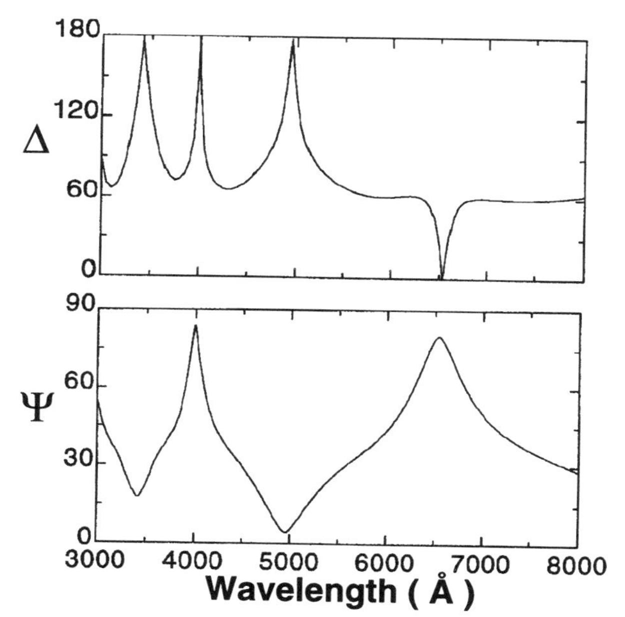

Figure 7.5 The ellipsometric spectra for a single film on a substrate with conditions similar to those in Figure 7.4 except that the values of the optical constants are those of silicon dioxide on silicon.

180° line rather than toward the zero line for Figure 7.4. This is due to the fact that the index of refraction of the substrate is significantly higher at 4000 Å for the model used for Figure 7.5 than for the model used for Figure 7.4, whereas the index at 6000 Å is about the same for the two models. This difference is analogous to the difference in features B and C in Figure 7.3. Note also that the other two cusps at the 180° line (analogous to feature A in Figure 7.3) are at slightly different spectral locations in Figure 7.5 than in Figure 7.4. These are also simply perturbations due to differences in optical constants.

7.7 CONTRIBUTION OF FILM THICKNESS

Figure 7.6 shows the ellipsometric spectra for oxide on silicon where the thicknesses of the oxide are 4500 Å and 4800 Å. Note specifically that when the thickness increases, the various features of the spectrum move to the right (toward a higher wavelength). Another feature which is not particularly evident in Figure 7.6, but is shown in Figure 7.7, is that as the thickness increases, the cusps become closer together.

A reasonable approach when attempting to develop seed values for a regression analysis is to propose a thickness, calculate the model values, and compare them to the measured values. If the spacing between cusps is approximately

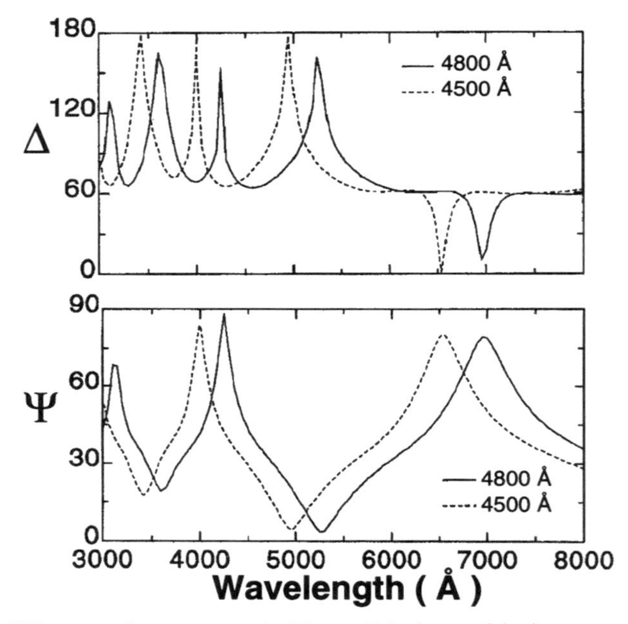

Figure 7.6 Ellipsometric spectra as in Figure 7.5 along with the spectra for a slightly thicker film. Note that the features appear to move toward higher wavelengths.

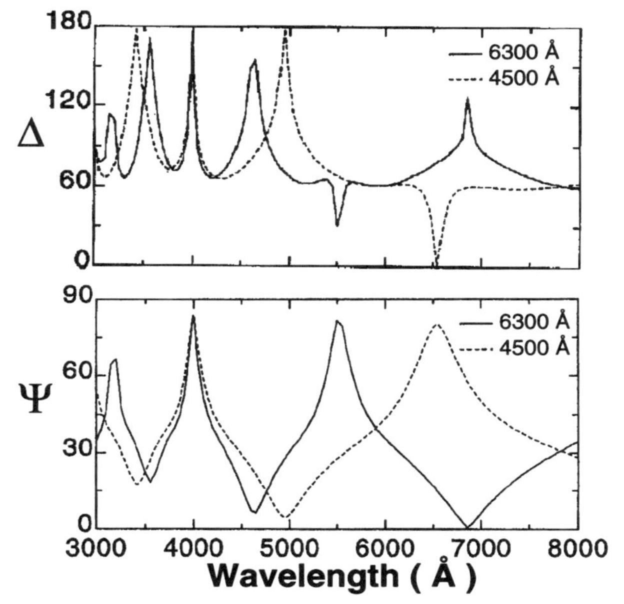

Figure 7.7 Ellipsometric spectra as in Figure 7.5 along with the spectra for a significantly thicker film. Note that although some of the features appear to match, most do not. This figure shows that for thicker films, the cusps are closer together.

correct but the model spectra is not at quite the right location, increase the model thickness by a small amount (e.g., 200 Å) to move the model thickness to the right, or decrease the thickness to move the model to the left until the model spectra almost matches the measured spectra. If the spacing between cusps in the model spectra is not particularly close to the spacing of the cusps in the measured spectra, add or subtract a greater amount of thickness (e.g., 1000 Å) to move the cusps a larger distance and to make them closer or further apart. When the model spectra roughly approximate the measured spectra, start the regression analysis to optimize the thickness and/or optical constants.

7.8 SUMMARY

In this chapter, we have discussed the look and feel of an ellipsometric spectrum for a single layer on a substrate. We have identified features which we describe as cusps, and have discussed the origin of these features. Variations due to wavelength change are related to variations due to thickness change. However, the variations due to wavelength change have the added complication that the optical constants also vary with wavelength, whereas the optical constants do not vary with thickness (in a homogeneous sample). Finally, we have shown that increasing the model thickness moves the features to a higher wavelength and causes the cusps to become closer together and decreasing the model thickness moves the features to a lower wavelength and causes the cusps to be farther apart.

7.9 REFERENCES

1. D. E. Aspnes, in *Handbook of Optical Constants of Solids*, edited by E. D. Palik, Academic Press, New York (1985), p. 90.
2. H. G. Tompkins, *A User's Guide to Ellipsometry*, Academic Press, New York, 1993.
3. R. M. A. Azzam and N. M. Bashara, *Ellipsometry and Polarized Light*, North-Holland, Amsterdam, 1977.
4. E. D. Palik (editor), *Handbook of Optical Constants of Solids*, Academic Press, New York, 1985.

CHAPTER 8

Analytical Methods and Approach

8.1 GENERAL APPROACH

The general approach for both spectroscopic ellipsometry and spectroscopic reflectometry is:

1. Ensure that the sample is positioned on the instrument correctly.
2. Collect a set of data which is sufficient to provide the required information.
3. Determine, in some manner or other, values of thickness and optical constants of all of the layers and the substrate in such a way that the agreement between the calculated data and the measured data is acceptable.

The first two items will be discussed briefly. Most of this chapter will deal with the analysis of the data suggested in Item 3.

8.2 SAMPLE POSITIONING

There are two issues in sample positioning for a reflection-based instrument. Although we illustrate these with an example of an ellipsometer, the issues are similar for a reflectometer, although for many reflectometry instruments, the angle of incidence is zero (normal incidence) and the positioning issue is taken care of in the construction of the instrument. For ellipsometry, the angle of incidence is one of the variables and is usually in the range 50–80°. Accordingly, the positioning issue must be dealt with for each sample.

For ellipsometry, the sample must be oriented (tilted) in such a way that the angle of incidence (and angle of reflection) are equal to the prescribed values, as shown in Figure 8.1A. Following this, the sample must be translated, without changing the tilt, so that the reflected beam will enter the instrument at the correct location and going in the correct direction, as shown in Figure 8.1B.

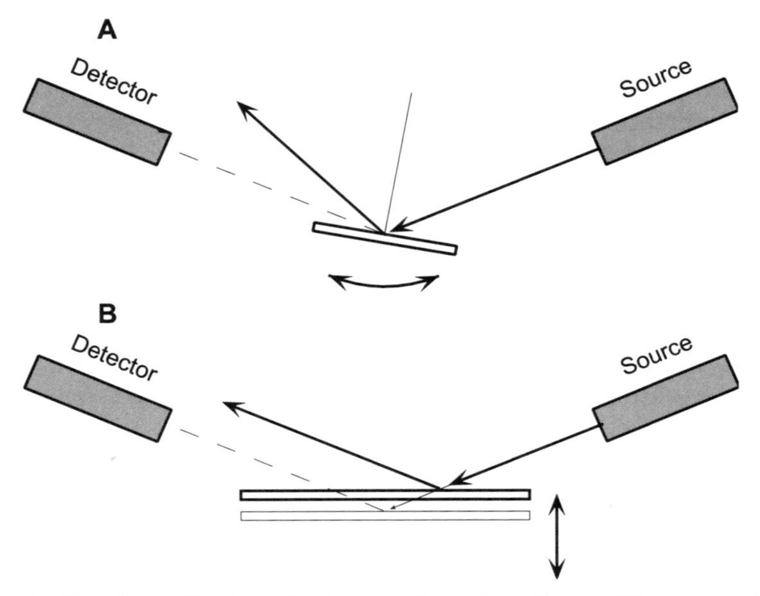

Figure 8.1 Sample positioning adjustments, shown here for an ellipsometer. (A) The tilt adjustment; (B) the translation adjustment.

The tilt adjustment sometimes uses an auxiliary light source and autocollimator, and the sample is tilted until the center of two sets of cross-hairs coincide. In other cases, the incident arm of the instrument is moved to an angle of incidence of zero, and white light from the regular instrument source is used along with an auxiliary detector.

Following the tilt adjustment, the translation adjustment is made either by visually observing when the reflected light is centered on an iris on the detector side, or by using the detector to determine when maximum signal is observed. It should be noted that the methods for sample positioning for spectroscopic ellipsometry are very similar to the methods used for single-wavelength ellipsometry.

8.3 HOW MUCH DATA?

8.3.1 Wavelength

Although plots of spectroscopic data are normally shown as a continuous line, in practice the data are taken at discrete wavelengths separated by some wavelength interval. In addition, some instruments have a wavelength range in the visible region only, whereas the range of some instruments includes the UV and/or the IR regions.

For materials with optical constants which vary slowly and smoothly with wavelength, the controlling factor for the wavelength (or photon energy) interval is the thickness. Figure 8.2 shows the plot for some of the ellipsometric data for a 4 μm layer of SiO_2 on a silicon wafer. The solid line shows the data when the wavelength interval is sufficiently small to delineate the spectrum well. The dashed line shows an example of how the spectrum might look if the interval was too large to delineate the spectrum properly. If the structure were very well known and the thicknesses were known approximately, even a few data points taken at large intervals would be sufficient to obtain an accurate thickness value (having 10 values of Δ is better than having only one, as is the case in single-wavelength ellipsometry). However, when the sample structure is unknown, or the thickness is not known even approximately, having an inadequately delineated spectrum is a severe handicap.

In some cases, we use ellipsometry to determine the optical constants of a substrate and draw conclusions from the structure of the optical constants. Figure 8.3 shows the values[1] of the index for $Al_xGa_{1-x}As$ for values of x of 0.3 and 0.7. The location of the cusp is important, and data points must be taken with a small enough increment in wavelength to delineate the index spectrum adequately.

8.3.2 Angle of Incidence

For reflectometry, the angle of incidence is normally not one of the adjustable parameters. For reflectometers found in a wafer fab, the angle of incidence is usually normal. For UV–VIS spectrophotometers sometimes found in analytical laboratories, angles up to 30° may be used. For ellipsometers, the angle-of-incidence should be chosen so as to maximize the difference between the intensities of the p-wave and the s-wave. In order to obtain this condition, one should use an

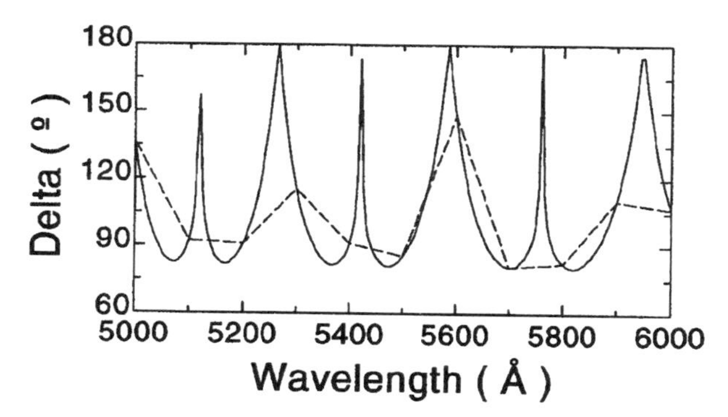

Figure 8.2 The effect of too large a wavelength interval. The sample structure is a 4 μm SiO_2 layer on a silicon substrate. The dashed line is the spectrum observed if a wavelength interval of 100 Å is used, and the solid line is the spectrum observed for the same structure when a wavelength interval of 5 Å is used.

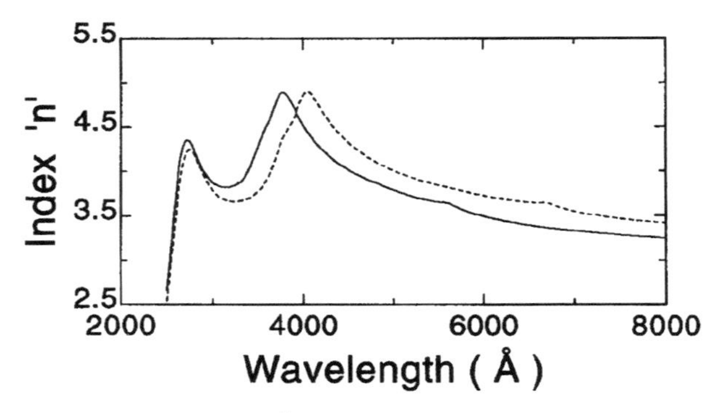

Figure 8.3 The index of refraction[1] of $Al_xGa_{1-x}As$ for values of $x = 0.3$ and $x = 0.7$. In order to use the index spectrum for stoichiometry determinations, one must take data at sufficiently small wavelength intervals to delineate the index spectrum.

angle which is near (but not exactly at) the Brewster angle for the substrate. This is particularly important for very thin films.

When determining the optical constants of an unknown material, it is often beneficial to obtain the data at two or three different angles of incidence, with at least one angle being above the Brewster angle and at least one below the Brewster angle. This is also helpful for multilayer stacks. Data may be taken at 5° or 10° increments. When the optical constants of the materials are well understood and only the thicknesses of the various layers are to be determined, a single angle of incidence may be used. Because the Brewster angle for single-crystal silicon in the visible region is about 76°, the angle of incidence for SE is often chosen to be 75° (for SWE, 70° is often the instrument manufacturer's choice). For glass substrates, the Brewster angle will be about 56°, and hence a choice of 60° for the angle of incidence might be more appropriate. When the layers are several thousand angstroms thick, the sensitivity to being close to the Brewster angle is reduced.

8.4 ANALYTICAL METHODS

8.4.1 General

After the ellipsometry or reflectometry data have been collected, it is necessary to use these data to determine the quantities of interest, such as thickness and optical constants. For a bulk material with no films, the optical constants may, in some instances, be calculated directly from the measured data. Tompkins[2] gives the necessary equations to calculate n and k from the ellipsometric parameters Δ and Ψ for a substrate. Reflectance data can be used for calculating the index of refraction of dielectrics, particularly if one is willing to assume a dispersion relationship such as the Cauchy relationship.

For more complex samples (e.g., single or multiple layers on a substrate), the quantities are not calculated directly. A model is postulated, with values of thickness for the layers and values of the optical constants for the layers and the substrate. The expected values of either the reflectance $\Re$ or the ellipsometric parameters Δ and Ψ are calculated and compared with the measured quantities. A regression analysis then refines the choices of values for the thickness and optical constants until the difference between the calculated values and the measured values is minimized. This process uses a merit function of some sort to measure the goodness of fit. The merit function which we shall use in this and subsequent chapters is called the mean-squared error, or MSE, and is given in detail in Chapter 9. Larger values of the MSE imply a worse fit. A better description might be the "badness of fit."

The values of the thickness and the optical constants in the original postulation are called the *seed values*. The seed values for the thickness typically come from the fabricator's estimate, based on the deposition parameters. The optical constants are handled in several different ways. In the following sections, we discuss some of the ways of using or determining the optical constants.

8.4.2 Tabulated Optical Constants

Tabulated values of optical constants can be found in several places. Examples of these are the two-volume set edited by E. D. Palik[3], and *The American Institute of Physics Handbook*[4]. Another example is where one has previously determined the values (see the section on point-by-point fit, below) for a particular material and stored the values in the computer.

In these cases, the values of n and k are listed at various values of wavelength. These values can often be used directly for materials which can be fabricated in a very reproducible manner. These include single-crystal silicon, GaAs, and in general any single-crystal material. Other materials such as the thermal oxide of silicon and LPCVD silicon nitride are also reproducible enough that the tabulated values can often be used directly. In this case, the regression analysis determines only the thickness (the tabulated optical constants are used and therefore do not have to be determined). Software on most spectroscopic ellipsometers and the newer reflectometers will usually have the ability to access tabulated optical constants which are stored as ASCII text files in the computer. This is the simplest form of analysis, with the classic example being thermal oxide on a silicon wafer.

For many materials, however, the fabrication of the materials is not reproducible enough to use the tabulated values directly. Examples of these materials are polycrystalline metals, PECVD nitrides, oxides, oxynitrides, indium–tin oxide, and so on. In these cases, the tabulated values are not accurate enough for direct use, but can serve as seed values for additional refinement in the values for the material of interest. The refinement is done during the regression analysis.

8.4.3 Simple Mixtures of Tabulated Optical Constants

In some cases, the optical constants of a material of interest can be adequately described by mixing the optical constants of two known materials (materials for which tabulated values are available). The mixture is usually done using the *effective medium approximation* (EMA). A thorough discussion of the EMA is given by Aspnes et al.[5] and summarized in an appendix to *A User's Guide to Ellipsometry*.[2] The Maxwell Garnet approximation and the Lorentz–Lorenz approximation are effective medium theories that represent heterogenous mixtures. These two approximations work well when the mixture consists of a small amount of one material in a matrix of the other material. The Bruggeman EMA extends this to include materials which span the entire range of composition. The software included with most ellipsometry instruments and some reflectometry instruments can do this calculation readily given the fractional components. In this case, in addition to the thickness of layers, the composition fraction is also one of the parameters which are determined.

One example of this might be spin-on-glass on a silicon wafer. The spin-on-glass is basically silicon dioxide, although the density is not that of silicon dioxide formed by thermal oxidation. The optical constants of this material might be generated by an EMA mixture of thermal oxide and a few percent of *voids*. The use of voids, with optical constants of empty space, is simply a convenient way to raise or lower the values of a known material to obtain the values for the material of interest. The term should NOT be taken to imply the presence of small cavities which could be seen with high-resolution microscopy. For a material which has optical constants which are slightly higher than those of the known material, the mixture might include *negative voids*. Again, the term voids should not be taken literally.

Another example of an EMA mixture might be an oxynitride. Figure 8.4 shows the measured spectra of a PECVD silicon oxynitride (dashed line). The solid line in (A) represents the best fit if one uses the tabulated optical constants of silicon dioxide, giving a thickness of 2904 Å with an MSE of 228. The fit is obviously not very good, and the MSE indicates this. The solid line in (B) represents the best fit if one uses the tabulated optical constants for LPCVD silicon nitride, giving a thickness of 1766 Å with an MSE of 283. Again, not a good fit.

If one uses an EMA mixture of the optical constants of the oxide and nitride, and allows the concentration fraction to be one of the regression parameters, one obtains the solid line shown in (C). In this case, the nitride concentration is 41.2%, the thickness is 2290 Å and the MSE is 20, indicating a reasonably good fit.

Films having nitride fractions up to about 70% could be described rather well with an EMA approximation, but films having nitride fractions of greater than this could not adequately be described by this simple mixture, primarily because of the presence of excess silicon, which cannot be modeled with a mixture of oxide and nitride.

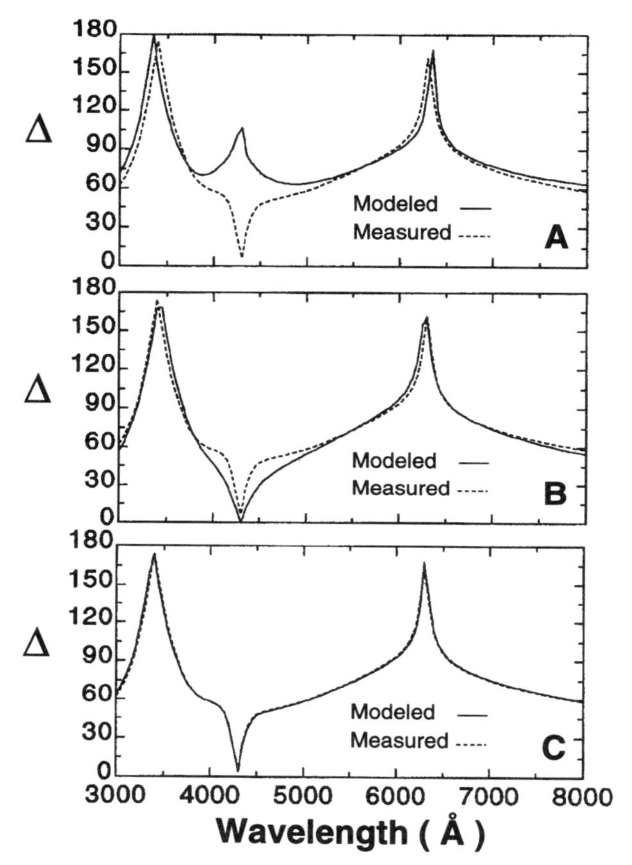

Figure 8.4 Measured and modeled values of delta for a PECVD silicon oxynitride film. (A) Modeled as an oxide; (B) modeled as a nitride; (C) modeled as an EMA mixture of oxide and nitride. See text for details.

LPCVD polycrystalline silicon (polysilicon) is often described by an EMA mixture of crystalline silicon along with about 15% amorphous Si and about 5% voids. Figure 8.5 shows measured data along with the fit. Although the fit is very good for most of the spectral range, there is some misfit at wavelengths of the critical points of crystalline silicon. We shall see that a slightly better fit can be obtained using a dispersion relationship (described below).

Another use of the EMA model is to describe a roughness layer. All of the calculations for ellipsometry assume plane parallel interfaces. If a substrate is microscopically rough,[6] the material can be described, as shown in Figure 8.6, as a perfect substrate separated from the perfect ambient by an effective medium layer (the roughness layer). The optical constants of the effective medium layer are obtained by mixing the optical constants of the material and the optical constants of the ambient. Normally the thickness of this layer is strongly cor-

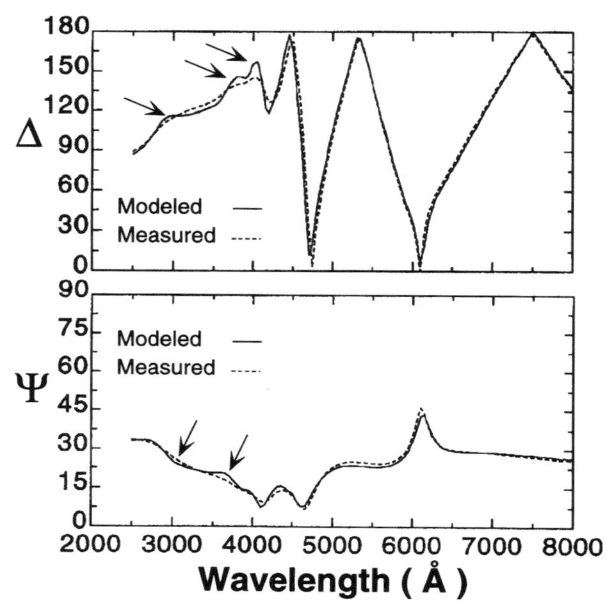

Figure 8.5 Measured and modeled values of delta and psi for polysilicon on silicon dioxide on a silicon wafer. The model for polysilicon is an EMA mixture of crystalline silicon with 17.4% amorphous silicon and 4.2% voids, with thickness of 1626 Å. The model also includes a roughness layer (see text) of 24 Å, and the oxide layer thickness is 1176 Å. The MSE is about 90, which is an acceptable fit, but not exceptionally good.

related to the composition and hence cannot be determined independently. For convenience, the composition is often simply taken as half material and half air. This approach also works well for a roughness layer between two solid materials. Roughness is discussed at length in a subsequent chapter.

The EMA model is used primarily for simplicity. Note that the number of regression parameters is still quite small (thickness and one or two fractional compositions per layer) compared with a rather large number of measured values (one per wavelength for reflectometry and two per wavelength for ellipsometry).

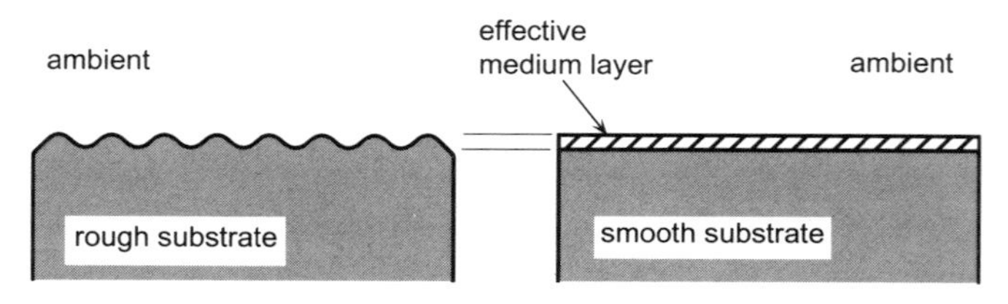

Figure 8.6 A substrate which has microscopic roughness can be modeled as a perfect substrate separated from a perfect ambient by an effective medium layer.

A low value of the MSE implies a reasonable representation of the optical constants and thickness. One must be cautious, however, in drawing conclusions about the material concentration fraction values, since this is based on an assumed model. In the case of the oxynitride mentioned above, Rutherford backscattering spectrometry and Fourier transform infrared spectroscopy shows that the oxide fraction is overestimated by SE, using the EMA model as presented (H. G. Tompkins, unpublished results). This is due to the fact that we have ignored the hydrogen which is always included in PECVD oxynitride films.

8.4.4 Dispersion Relationships

The next level of complexity is to use a more sophisticated dispersion relationship to describe the optical constants of a material. Although the tabulated optical constants are simple values for each wavelength, the values are not independent of each other. The values of n and k should vary smoothly as one varies the wavelength in a continuous manner. Dispersion relationships are simply equations which give the values of n and k as a function of wavelength. Although some of the possible relationships are strictly empirical, most are based to some extent on physical principles.

The term *dispersion* was originally used[7] to describe how far a prism of a material will separate the extreme colors (detected by eye) from the *mean ray*. The white light spectrum projected onto a screen will be physically larger with a prism of greater dispersive power than one with a lesser dispersive power. This is determined by the difference in the index n for the prism at the extreme wavelengths.

The best-known dispersion relationship was developed by L. Cauchy.[8] He undoubtedly observed that for most transparent material, the index n decreased with increasing wavelength (in the visible range) and could be expressed approximately as

$$n(\lambda) = n_0 + \frac{n_1}{\lambda^2} \tag{8.1}$$

where n_0 and n_1 are parameters.[9] n_0 gives the constant value at long wavelengths, and n_1 allows the value of n to increase for shorter wavelengths. An additional term is often added to give

$$n(\lambda) = n_0 + \frac{n_1}{\lambda^2} + \frac{n_2}{\lambda^4} \tag{8.2}$$

where n_0, n_1, and n_2 are known as the *Cauchy Coefficients*. n_1 controls the curvature in the middle of the visible spectrum and n_2 influences the spectrum to a greater extent for shorter wavelengths. Although one can rationalize this expression from physical principles, it is usually used in an empirical manner.

Including an additional term with an exponent of 6 allows even greater curve-fitting capability.

The extinction coefficient, k, is sometimes given as an expression similar to Eq. 8.2, and the coefficients are called the *Cauchy extinction coefficients*, although Cauchy is not the source of this relationship for k. Other relationships for k might use exponents of 1 and 3 rather than 2 and 4. One other relationship for the extinction coefficient is the Urbach[10] relationship

$$k(E) = A_k \, e^{B_k(E-E_b)} \tag{8.3}$$

where A_k, B_k, and E_b are the parameters. In this equation, we express k as a function of the photon energy E rather than the wavelength. Recall that they are related by

$$E \cong \frac{12\,400}{\lambda} \tag{8.4}$$

where E is given in electron volts and λ is given in angstroms.

It should be noted that Eq. 8.3 contains only two independent parameters. It can be shown that there are an infinite number of values of the couple (E_b, A_k) which will give the same result.

For tabulated optical constants one chooses the tabulated set, and there are no further choices to be made for optical constants. The EMA method gives some added flexibility (and complexity), but the number of regression parameters is still limited to one or two for a given layer. When one is using a dispersion relationship, one has from two to six parameters (or possibly a few more, depending on the choice of dispersion relationship) for the optical constants. This adds both complexity and flexibility. Note that for these examples of dispersion relationships, n and k are independent of each other and must both be fit in a regression analysis (although for dielectrics, k is often taken to be zero for the spectral range of interest). For other more sophisticated dispersion relationships, n and k are related to each other.

Another dispersion relationship which is used with some regularity is the oscillator. We discussed the Lorentz oscillator in Chapter 3, and some regression software for both ellipsometry and reflectometry will have the capability of using some form of combinations of oscillators for the dispersion relationship. In some cases, the relationship is based on physical principles and in other cases it is strictly empirical. The oscillator formalism uses complex numbers, and a given equation will include both n and k (or ε_1 and ε_2). Normally several oscillators are required, and hence the number of regression parameters may exceed 15 or 20.

Figure 8.5 showed the measured and modeled values for Δ and Ψ for polysilicon where the modeling was done with an EMA mixture. Figure 8.7 shows the same measured values along with modeled values using the Lorentz oscillator dispersion relationship. Note the significantly better fit in the UV part of the spectrum. The modeling was done using four Lorentz oscillators, with 13

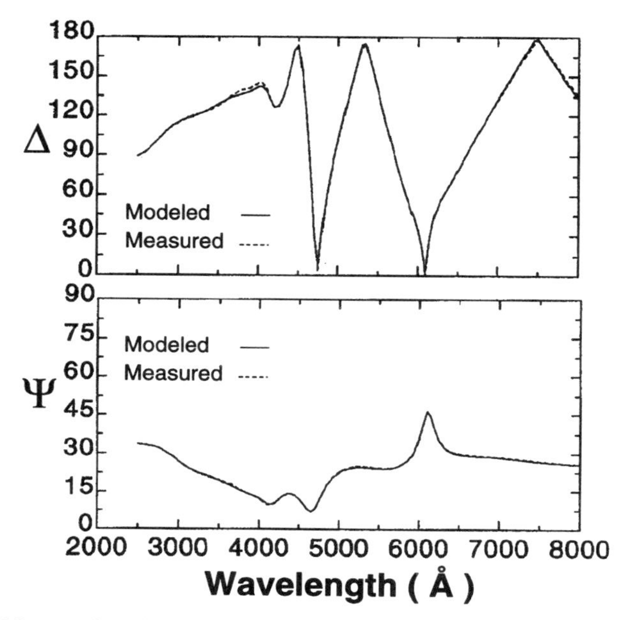

Figure 8.7 Measured and modeled values of delta and psi for polysilicon on silicon dioxide on a silicon wafer. The model for polysilicon is a Lorentz oscillator dispersion relationship with four oscillators, with thickness of 1634 Å. The model also includes a roughness layer of 26 Å, and the oxide layer thickness is 1150 Å. The MSE is about 17, which is a considerably better fit than that shown in Figure 8.5.

parameters for the optical constants. In this and all regression analyses, one must weigh the benefit gained (goodness of fit) against the added cost of complication (more regression parameters).

Other more complicated dispersion relationships are also possible, depending on the software capability of the instrument. The complexity of these is beyond the scope of this book, however, and they will not be dealt with here.

8.4.5 Point-by-Point Optical Constants

For some analyses, we ignore the requirement that the optical constants at various wavelengths should be related to each other. Suppose one has made measurements at M different wavelengths. One has $2M$ ellipsometric values and can calculate the $2M$ unknowns (n and k at each wavelength). If a film of unknown thickness is present, we have one more unknown than measured values (the thickness is not a function of the wavelength) and hence the thickness of the film and the (independent) optical constants of the film cannot be determined. This situation can readily be dealt with if one has several samples with films which have the same optical constants but different thicknesses.

One drawback of a point-by-point analysis is that the regression analysis is lengthy since one has hundreds of regression parameters. A second drawback is that experimental measurement noise in the measured data will be directly translated into noise on the optical constant spectra. The point-by-point analysis does not work well on production systems owing to a combination of bandwidth and angle-of-incidence range effects. With a system used in an analytical environment, it is sometimes useful to make a point-by-point analysis and then to approximate the resulting point-by-point optical constants with an oscillator dispersion relationship.

8.4.6 Regression Process

Regression analysis is discussed extensively in the next chapter. We summarize some of the salient features here. The regression software should have some method of building the model and entering the seed values for the regression parameters. The form of optical constants (tabular, dispersion relation, etc.) will usually be one of the inputs for building the model, although this is sometimes done implicitly in choosing the various layers for the structure. The initial seed values for thickness often are chosen from processing target thicknesses.

After building the model and inputting the seed values, we usually ask the software to generate the data for the initial model. The purpose of this step is to see if we are anywhere close to a reasonable fit and to estimate whether the regression process will find the global minimum in the misfit or will find a local minimum. If the fit seems reasonable, we progress to the next step. If the fit seems totally unreasonable, we must choose different seed values for the thicknesses. By trial and error, one can usually obtain generated values which are not too different from the measured values.

When the fit seems reasonable, the regression process is started. For a complex model, we often allow only a few regression parameters to vary. For example, if we feel that the optical constant model is reasonably good, we might allow the thicknesses of the layers to vary while holding the optical constant parameters fixed. After the best fit for thicknesses is found, we then allow the optical constant parameters to vary along with the thicknesses, to further optimize the fit. In some cases, we initially fit for n while keeping k equal to zero. If a good fit is obtained in the visible and IR part of the spectrum, but the UV part of the spectrum does not fit, allowing nonzero values of k will further optimize the fit.

In some cases, the fit can be improved by adding in additional layers which represent roughness between materials or interdiffusion. Some software allows for the possibility of a graded layer. In general, one must decide whether the improved fit justifies the increased complexity. The "best" answer is the simplest one which gives a reasonable fit.

8.5 SUMMARY

With a properly operating instrument, we position the sample and collect data at the appropriate angles of incidence and at appropriate wavelength intervals. We then build a model of the structure with seed values for the thickness and the optical constants for each layer. Using the regression analysis, we determine the final values for thickness and the optical constants which give the best fit between the calculated values and the measured values.

8.6 REFERENCES

1. The data for this plot was taken from software on the J. A. Woollam Co. VASE® instrument.
2. H. G. Tompkins, *A User's Guide to Ellipsometry*, Academic Press, New York, 1993.
3. E. D. Palik (editor) *Handbook of Optical Constants of Solids*, Academic Press, New York, 1985; E. D. Palik (editor), *Handbook of Optical Constants of Solids II*, Academic Press, New York, 1991.
4. D. E. Gray (coordinator editor), *American Institute of Physics Handbook*, 3rd ed., McGraw Hill, New York, 1972.
5 D. E. Aspnes, J. B. Theeten, and F. Hottier, *Phys. Rev. B*, **20**, 3292 (1979).
6. H. G. Tompkins, *A User's Guide to Ellipsometry,* Academic Press, New York, 1993, Chapter 7.
7. D. Brewster, *Optics*, Longman, Rees, Orme, Brown, and Green, London, 1831, p. 75.
8. L. Cauchy, *Bull. des. sc. math.*, **14**, 9 (1830).
9. J. Strong, *Concepts of Classical Optics*, Freeman, San Francisco, 1958, p. 69.
10. F. Urbach, *Phys. Rev.*, **92**, 1324, 1953.

CHAPTER 9

Optical Data Analysis

9.1 INTRODUCTION

Optical characterization techniques such as reflectance and ellipsometry are nearly always indirect measurements. It is simply not possible to measure a film thickness or optical constant directly with light; instead we measure the reflectance or ellipsometric parameters of a given sample. The measured data are then modeled in order to determine the sample properties of interest (film thicknesses, optical constants). An exception to this case is the direct measurement of the absolute reflectance of a sample (for example to gauge the roughness of an aluminum film) where the reflectance alone is of interest.

As a result, there are three steps to any optical measurement. First, the optical data (reflectance or ellipsometric parameters) must be measured. Second, the measured data must be modeled in order to determine the sample parameters of interest. Finally, the results of the modeling procedure must be gauged to estimate the accuracy and precision of the result. All three stages are very important in any application, whether it is production or research related. It is also important to understand that errors can be introduced by both the first and the second stages. The measured reflectance or ellipsometric data may contain both random and systematic errors due to the measurement system. The modeling process will induce further systematic errors if the optical model is not accurate for the given sample. Finally, if the model fit is not good, or worse, is not unique, the final result can be completely wrong, even if the original measured data is accurate and noise free.

In this chapter we discuss the techniques used to extract useful information from measured reflectance and ellipsometric data. Particular attention is paid to sources of errors and the analysis of these errors in the data. The chapter concludes with a general troubleshooting section for debugging optical characterization problems.

9.2 DIRECT CALCULATION

There are two basic ways to approach the analysis of optical data. The first method is to calculate the parameters of interest in the optical model directly from the measured optical data. The second approach is to vary the parameters of interest in the model in some systematic way in order to get the calculated model data to match the measured data as closely as possible. This type of approach is termed regression analysis, and is discussed in the next section.

Direct calculation methods may be applied to both reflectance and ellipsometric data. One of the most well-known direct calculation methods is the inversion of ellipsometric Δ and Ψ data to obtain the optical constants of an unknown material. Ellipsometry measures two parameters, Δ and Ψ, at any given wavelength. As a result, it is possible to obtain any two sample parameters by directly inverting the ellipsometric equation

$$\tan \Psi e^{j\Delta} = \frac{r^{p}}{r^{s}} \tag{9.1}$$

where r^{p} and r^{s} are the complex Fresnel reflection coefficients for the substrate. These equations can be inverted to give n and k as a function of Δ, Ψ, and the angle of incidence. (This is discussed briefly in Chapter 6.) In principle, one should be able to extract any two unknown quantities from a film-covered substrate where all of the other quantities are known (e.g., thickness and index of refraction of a transparent film). In practice, however, the equations cannot be inverted and a regression analysis, discussed later in this chapter, must be used.

There are a number of analytical techniques which are used to extract sample information from reflectance spectra. Of particular importance are those which analyze the location of extrema in the reflectance spectrum in order to obtain information about the thickness and index of refraction of transparent films. A typical reflectance spectrum from a transparent film is shown in Figure 9.1.

Note the oscillation of the reflectance data as a function of wavelength. This is due to constructive and destructive interference between the light reflected from the top and bottom surfaces of the oxide film. For normal incidence and for a film with an index of refraction less than the index of the substrate, if two adjacent maxima in the reflectance spectrum are located at wavelengths λ_1 and λ_2, the thickness of the film may be directly obtained[1] from

$$n_1 d_1 = \frac{\lambda_1 \lambda_2}{\lambda_1 - \lambda_2} \tag{9.2}$$

where n_1 is the index of refraction of the film, d_1 is the thickness of the film, and λ_1 is assumed to be greater than λ_2. This equation provides a very useful means

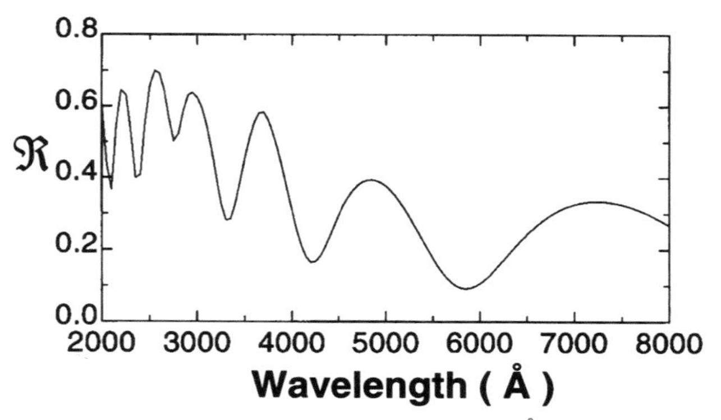

Figure 9.1 Absolute reflectance calculated for a 5000-Å film of silicon dioxide on silicon.

for quickly obtaining the thickness of a single transparent film from reflectance data, given its index of refraction. Note that for single transparent films in general, the distance between adjacent maxima in the reflectance data, and hence the period of the interference oscillations in the reflectance data, is proportional to the product of the film thickness and index of refraction. The amplitude of the interference oscillations is independent of the thickness of the film (provided the film is thick enough to exhibit interference oscillations), but is strongly dependent on the difference in index of refraction between the film and the substrate. This was discussed at length in Chapter 5.

Thick transparent films generally cause many interference oscillations to appear in the reflectance data. A simple way to extract thickness information from structures containing one or more thick transparent films is to Fourier analyze the experimental reflectance data, with the reflectance plotted versus the photon energy (rather than the wavelength). If the dispersion of the film index of refraction is ignored so that the film index is assumed to be constant over some spectral range, the interference oscillations due to the film in that spectral range will have a period which is constant. A Fourier analysis of such data yields a peak at a single frequency. This frequency is proportional to the optical thickness of the film (physical thickness multiplied by the film index of refraction). This technique is particularly useful for very thick films, and has also been applied to the estimation of multiple layer thicknesses in film stacks.[2]

These analytical techniques can be very useful for simple structures such as single films, but tend to be of limited use for more complicated structures. When more than one or two parameters are to be obtained from the measured optical data, it is usually necessary to employ more general regression techniques to analyze the measured data. In spite of this, there are a large number of commercial optical metrology tools (reflectometers and ellipsometers) which employ direct calculation techniques for the analysis of the measured data.

9.3 REGRESSION ANALYSIS METHODS

9.3.1 The Problem

It is often not possible to extract the thicknesses or optical constants of interest directly from the measured optical data, particularly if more than one or two parameters are to be obtained from the analysis. As an alternative to direct calculation methods, regression methods may be used to extract the sample parameters of interest from the measured optical data.

The problem of optical data analysis can be stated as follows. We have an experimentally measured set of optical data from a given sample. This data set may consist of any combination of reflectance, transmission, ellipsometric, or other optical data types. We also have a model for the sample which allows us to calculate the reflectance, transmission, or ellipsometric data for the model given the thicknesses of all layers in the model and the optical constants of all layers and the substrate in the model. We wish to adjust the values of some of the parameters in the model (thicknesses and optical constants) so that the data calculated from the model matches the measured data as closely as possible. To do this, we need two things. First we require some sort of merit function which indicates how well the experimentally measured data match the calculated model data. Second, we require an algorithm which can adjust the desired model parameters in order to optimize this merit function.

9.3.2 The Merit Function

Merit functions which are used to judge the quality of the match between the measured and calculated data are known as maximum likelihood estimators. There are many possible choices for a maximum likelihood estimator, but the most common is the mean-squared error (MSE):

$$\mathrm{MSE} = \frac{1}{N-M}\sum_{1}^{N}\left(\frac{y_i - y(\bar{x}, \bar{a})}{\sigma_i}\right)^2 \qquad (9.3)$$

In this equation, y_i denotes the experimentally measured data values, $y(\bar{x}, \bar{a})$ are the model calculated data points, and σ_i is the standard deviation of the ith data point. The vector $\bar{x}$ contains all known parameters in the model, while the vector $\bar{a}$ contains the model parameters which are to be varied in order to fit the experimental data. There are N total data points, and a total of M variable parameters. This function will exhibit a minimum value of zero when the calculated model data exactly match the experimentally measured data. In practice, the model is usually unable to match the experimental data perfectly, and in that case the MSE will exhibit a minimum value when the model matches the experimental data as closely as possible.

Note that in this formalism, the contribution for each data point to the total MSE is divided by the square of the standard deviation of that data point. This provides a physical interpretation for the actual value of the MSE. When the MSE equals one, the calculated data points are within one standard deviation of the experimental data points (on the average over the entire data set). The weighting of the individual contributions of each data point by its standard deviation also prevents noisy data points from skewing the results of the fit. Noisy data points will generally have large standard deviation values, so that their contribution to the MSE tends to be smaller. The MSE in this form would be considered a biased estimator.

Occasionally practitioners assume a value of unity for the standard deviations on all experimental data points. In this case, the physical interpretation of the MSE is much more difficult and all data points (noisy or not) are equally weighted in the data fitting, giving an unbiased estimator. Jellison[3] discusses the merits of the biased estimator versus the unbiased estimator, and Press et al.[4] also discuss estimators.

9.3.3 The Regression Algorithm[4]

At this point we have a set of experimental data points, a model function which can be used to calculate data points corresponding to the measured ones, a merit function (the MSE) which describes how well the calculated model data match the experimental data, and a set of variable parameters in the model which we want to adjust so that the calculated data match the experimental data as closely as possible.

The next step is to define an algorithm which can be used to adjust the values of the variable parameters in the model. The goal of this algorithm is to get the calculated model data to match the experimental data as closely as possible, giving a global minimum for the MSE value. There are a number of algorithms in the literature to perform this minimization. A few of the most popular ones are discussed in Appendix A.

The regression algorithm will iteratively try various values of the regression parameters until a minimum value for the MSE has been reached. Ideally, this will be a global minimum, although in many cases, a local minimum is reached. One way to help ensure that the minima is global is as follows. We use the thickness of a layer as the sample parameter

- Let z_1 be the seed value which you choose for the thickness variable z.
- Allow the regression to determine the minimum MSE and the corresponding value z_0 for the variable.
- Do a sanity check by visually comparing the measured and modeled spectra. If the fit is totally unreasonable, add or subtract about 1000 Å to the seed value and start again.

- If the fit is close, start the regression procedure over again, choosing a seed value z_2 which is a few hundred angstroms on the opposite side of z_0 from z_1. If the minimum value for the MSE is obtained when z is equal to the same value as z_0, the minimum is probably global.

9.4 HYBRID METHODS

Both the analytical and regressive types of algorithm described in the previous sections have relative strengths and weaknesses. Analytical algorithms are very fast, but tend to be rather limited in scope. Regression algorithms are very general by nature, but tend to be computationally slow if more than a few variable parameters are involved or if the data set to be modeled is large. Algorithms which combine both analytical and regressive techniques can take advantage of the strengths of both types of algorithms, and are very common among production-oriented optical metrology tools. Usually, analytical techniques are used to provide initial guesses at the values of the variable parameters in the optical model, and regression techniques are then used to fine-tune the model.

9.5 GOODNESS OF FIT

One of the common features of all techniques for modeling optical data is that they will always produce some sort of result. The user of these techniques is then left wondering whether or not this result is any good. There are a number of useful techniques for evaluating the quality of the results of a modeling procedure. The most obvious check on the fit quality is the value of the final MSE from the fit, particularly if real standard deviation values were used in the calculation of the MSE. Large values of the final MSE are usually indicative of a poor fit to the experimental data. It is also a good idea in general visually to compare the measured optical data and the final best-fit calculated data.

Most regression algorithms will also calculate the 90% confidence intervals on the variable parameters as a by-product of the analysis routine. These numbers are also very helpful for evaluating the quality of the best-fit result. If the 90% confidence interval for a given parameter is excessively large, then that parameter is either correlated to another model parameter or it does not effect the calculated data for the model. In this case it is a good idea to remove this parameter from the regression process.

The final and most useful test for the quality of a given data analysis is to ask whether the results are reasonable. Sure signs of failure are negative thickness values, zero thickness values, and nonphysical optical constant dispersion. For example, the index of refraction of transparent materials must always decrease with increasing wavelength. If an optical constant analysis of a transparent film yields index values which increase with increasing wavelength, the results cannot be correct. Most of the time nonphysical results from the modeling process are an indication that the optical model for the sample is not appropriate.

9.6 UNIQUENESS

As mentioned in the previous section, most modeling techniques tend to produce some sort of result which suggests whether or not the optical model and/or the measured data are accurate for the sample under test. In the previous section, methods for evaluating the quality of the fit to the experimental data were discussed. However, there is another, more subtle, problem which can be encountered in the analysis of optical data. Even though the quality of the fit to the experimental data might be good and the final best-fit results might be physically reasonable, it is possible that the solution generated by the modeling algorithm may not be unique. In other words, there may be more than one set of values for the variable model parameters which yields the same fit to the experimental data. This condition is referred to as *correlation*, and this will usually occur when there are too many variable parameters in the model, or when the effects of one parameter can exactly offset the effects of another parameter. This issue is discussed in detail by Press et al.[4]

A classical example of correlation is as follows:

- deposit a 300-Å layer of SiO_2 on a Si substrate, followed by a 500-Å layer of SiO_2;
- set up a model for the analysis consisting of two SiO_2 layers on a Si substrate with proposed seed values of 400 Å and 400 Å for the thicknesses of the two oxide layers;
- start the regression analysis software.

The regression analysis will probably find a solution which is very near the seed values and the fit will be extremely good. The correlation will be extremely high, however, since any combination of thicknesses which totaled 800 Å would have given an equally good fit. Changes in the thickness of one layer can be exactly offset by opposite changes in the thickness of the other layer.

We have used two thicknesses as the correlated parameters in this example. Correlation often exists between thicknesses and optical constants as well.

There are a number of useful tests which can be used to establish the uniqueness of a given solution. Most regression analysis software will calculate the sensitivity correlation matrix for the variable parameters. If the magnitude of any of the off-diagonal elements of this matrix approach unity then strong parameter correlations exist in the model. Also, large values of the 90% confidence intervals for two or more parameters in the model are often an indicator of strong parameter correlations.

Some regression software packages do not allow the analyst to view the correlation matrix or the 90% confidence intervals on the variable parameters after an analysis. In this case, the static repeatability of the given analysis can provide a good indication as to the uniqueness of the solution. Repeat the measurement and analysis of the data many times (10–20) and check the repeatability of the best-fit parameter values. Start from different seed values. If the results of the analysis show poor repeatability, it is very probable that there are strong parameter correlations in the model and that the best-fit parameter values are not unique.

If strong parameter correlations do exist in the optical model, there are two basic approaches to correcting the problem. First, additional optical data can be added to the data set. For example, if analysis of reflectance data alone does not yield a unique solution for a given set of parameters, the analysis of reflectance and ellipsometric data simultaneously for the given sample might do so. Second, the variable parameter set may be reduced. In general, the more variable parameters there are in the optical model, the less repeatable the analysis results tend to be. Reducing the number of variable parameters in the optical model tends to increase the repeatability of the analysis. It is particularly helpful to identify which parameters are strongly correlated by examining the correlation matrix and fixing the value of one or more of the strongly correlated parameters.

When multiple layers with unknown thicknesses are involved, one will often have correlation if the optical constants of some of the materials are also to be determined. Depositing single layers of the materials on a known substrate for the determination of the optical constants and/or measuring the thickness of underlying layers prior to depositing the upper layers will often reduce correlation significantly.

9.7 COMBINED ANALYSIS OF MULTIPLE DATA TYPES

The use of combined optical data types is a powerful means of reducing parameter correlations in the optical modeling process.[5] Simply stated, it is generally advantageous to amass as much independent experimental data about the sample under test as possible. Reflectance, transmittance, and ellipsometric data at different angles of incidence all provide additional information about the sample under test. It is important when analyzing more than one type of data simultaneously to make sure that all of the measurement systems used are reasonably accurate and that data were acquired from the same location on the sample under test.

9.8 MULTIPLE SAMPLE ANALYSIS

One of the most powerful practices for single-wavelength ellipsometry in determining optical constants is to use many different samples which have the same optical constants, but different thicknesses.[6,7] This is also true for spectroscopic ellipsometry and reflectometry.

As an alternative to the use of multiple optical data types, it is often possible to use multiple samples to amass experimental data concerning a particular material under study. For example, the determination of the thickness and optical constants of thin metal films is a very difficult problem owing to very strong correlations between the film thickness and the film optical constants. This correlation can be suppressed by simultaneously analyzing optical data from several metal films of differing thickness and assuming the optical constants of the different films to be identical. The data can then be analyzed to determine one set of optical constants in addition to the thickness of each film.

9.9 ERRORS—SYSTEMATIC AND RANDOM

Two basic types of errors are encountered in optical metrology—random errors, or measurement noise, and systematic errors. Each type of error has different effects on the data analysis procedure and the results of the data analysis.

Random noise on the measurement data can have a very strong detrimental effect on analytical data analysis techniques, but tends to have very little effect on regressive analysis techniques. Direct calculation techniques rely on the actual inversion of the measured data or the location of extrema in the measured data. Inversion techniques will carry noise in the measurement data directly into the results, while the location of extrema can be difficult if there is significant noise on the measured data.

If the noise on the measured data is truly randomly distributed, regression analysis of the data should still yield accurate results, as the noise will tend to average out over the entire data set. Also, if true standard deviations are used for the measured data in the calculation of the MSE, excessively noisy data points will be effectively excluded from the data analysis. In this case the noise will primarily affect the repeatability of the results, but will not strongly affect the accuracy.

Systematic errors can occur in both the measurement of the optical data and in the optical model. Examples of systematic errors in the measurement of the optical data would be errors in the angle of incidence or in the wavelength calibration of the measuring instrument. For the optical model, neglecting surface roughness of a rough film would be a typical systematic error, as would assuming a film to be transparent when it actually showed some optical absorption. It is impossible to detect systematic errors in either the measured data or the model by statistical analysis means. Systematic errors tend not to affect the repeatability of the measurements, but they have a strong effect on the accuracy of the measurement results.

9.10 REFERENCES

1. O. S. Heavens, *Optical Properties of Thin Solid Films*, Dover Publications, New York, 1965, p. 115.
2. M. Horie (of Dainippon Screen Mfg. Co., Ltd.), Method of measuring a thickness of a multilayered sample using ultraviolet light and light with wavelengths longer than ultraviolet, US Patent No. 5 440 141, August 8, 1995.
3. G. E. Jellison, Jr., *Appl. Opt.*, **30**, 3354 (1991).
4. A general reference for regression analysis is W. H. Press, B. P. Flannery, S. A. Teukolsky, and W. T. Vetterling, *Numerical Recipes*, Cambridge University Press, Cambridge, 1988.
5. B. Johs, R. H. French, F. D. Kalk, W. A. McGahan, and J. A. Woollam, *SPIE,* **2253**, 1098 (1944).
6. H. G. Tompkins and S. Lytle, *J. Appl. Phys.*, **64**, 3269 (1988); H. G. Tompkins, *Thin Solid Films*, **181**, 285 (1989); H. G. Tompkins, *J. Appl. Phys.*, **70**, 3876 (1991).
7. H. G. Tompkins, *A User's Guide to Ellipsometry*, Academic Press, New York, 1993, Chapter 4.

CHAPTER 10

Quality Assurance

10.1 GENERAL

In this chapter we discuss the methods and ideas for ensuring that the quality of the answer provided by these techniques is the highest possible. We first discuss some aspects of precision and accuracy, as applied to ellipsometry and reflectometry. Then we discuss some methods of determining whether the measuring instrument is operating at a prescribed performance level. This might take the form of routine measurements on a stable sample to ensure that the instrument is performing as well on a given day as it was the previous week.

10.2 PRECISION VS. ACCURACY

Precision is the quality of being able to obtain nearly the same answer when measuring an unchanging sample over and over again. If the variation of the answer is less than 1%, we would say that the technique is very precise, whereas if the variation of the answer is within plus or minus 10%, we would say that the technique is not particularly precise. Note that nowhere in this statement about precision is there any mention as to the closeness of the result to the truth.

Accuracy is the quality of being as close to "truth" as possible. For convenience, we do not yet define how one obtains truth in order to make the comparison. Many process engineers emphasize precision rather than accuracy, and precision ensures that what is being made today is very similar to what was being made previously. The importance of accuracy comes into play when a process is being transferred to another deposition system or when a different tool is used for measurement.

Suppose that over a 1-month period, one were to measure a 200-Å silicon nitride layer on a GaAs substrate daily with a single-wavelength ellipsometer. Further suppose that we determine the thickness using a model of a silicon nitride layer on a silicon wafer (rather than on a GaAs wafer). Since silicon nitride is very stable, if the sample were kept reasonably clean, one would repeatedly obtain a thickness of about 221 Å, suggesting that the measurement tool is very precise. The thickness value is not particularly accurate, since we are using an inappropriate model.

Ellipsometers and reflectometers are very precise, in that when they are operating properly, they have the ability to measure delta, psi, or reflectance very precisely (and very accurately). However, the quantity of interest to most process engineers is not these values, but the calculated values of thickness and/or optical constants. These determinations depend on having a model which is reasonable. This includes assumptions such as uniform layers with plane-parallel interfaces. It also includes a regression analysis which is supposed to determine when the deviation of the measured values from the calculated values is at a global minimum. The accuracy of the thickness values and optical constants therefore depends on how close the model is to reality, and whether the model structure is unique. Often there are several models which give equally good fits to the data.

Accordingly, for quantities such as thickness, optical constants, and so on, we can say that these techniques are very precise, but the accuracy depends on the ways in which the analyst deals with the data.

One should note that, except for measurements where one simply holds a yardstick next to the material and visually reads the thickness, ALL measuring techniques depend on a model, and hence the comments above also apply to all other measurements. It is not uncommon for a novice in the metrology business to compare values obtained from ellipsometry to values obtained from other measuring techniques (e.g., transmission electron microscopy) and to ask questions about accuracy of the optical techniques while totally ignoring how the quality of the other technique was ensured. This tendency should be resisted.

10.3 STANDARDS

The National Institute of Standards Technology (NIST) offers several standards (actually, standard reference materials) for use with ellipsometers.[1] These, and secondary standards derived from these are used routinely (often incorrectly) to reassure the operator that all is well. The series of standard reference materials (SRMs) which are thermal oxide on silicon have nominal thicknesses which range from about 2000 Å down to about 125 Å. The quantity which NIST certifies, however, is not the thickness, but the values of delta and psi at a wavelength of 6328 Å. The documentation gives the thickness when one uses a single-layer model, and also gives thicknesses when one uses a model of an oxide layer separated from the substrate by an intermediate layer. It is worth repeating that the thickness of the layer is *not* the certified quantity. Note also that the values of delta and psi are not specified or certified at any other wavelength except 6328 Å.

Proposals[2] have been made to use a clean single-crystal silicon wafer surface as a standard. Although most silicon surfaces have a small native oxide, when the silicon wafer is treated in a prescribed manner,[3] the silicon will be *hydrogen terminated* and will contain no native oxide. This type of sample is stable for several tens of minutes, but not several days or weeks. The recommendation is to use silicon ⟨111⟩ as opposed to silicon ⟨100⟩ owing to very slight roughening.

10.4 QUALITY ASSURANCE MEASUREMENTS

There are several types of routine measurements which can be and are made with ellipsometers and reflectometers to assure the operator that the instrument continues to operate properly. There is a tendency for process engineers to want to use samples with a structure which is similar or identical to the product. An example of this might be a polysilicon layer on silicon dioxide on a silicon wafer. Although this is appropriate for ensuring that the deposition process is in control, a simpler sample should be chosen to ensure the quality of the measurement technique. The choice of quality assurance samples should be such that the salient features of the *measuring equipment* is exercised and shown to be working properly rather than to show that the process is reproducible.

In order to put these measurements into perspective, let us consider the quantities which are important for the optical measuring methods. Ellipsometers measure the quantities delta and psi and reflectometers measure reflectance. Spectroscopic ellipsometry and reflectometers measure these quantities at various wavelengths. The angle of incidence is significant for ellipsometers.

For delta and psi, NIST standard reference materials exist[1] which have certified values of delta from about 90° to about 145°, and psi values from about 10.3° to about 45°. Unfortunately, there do not appear to be standard reference materials which have certified delta values between 90° and zero or near 180°.

Reflectance requires a quantitative measurement of light intensity. Since is it not normally convenient to measure the incident light directly, a reference sample is used, and the quantity which is then measured on the sample is the relative amount of light compared with the standard. Accordingly, the property of interest from a quality assurance point of view is the linearity of the detector. Since it is not particularly straightforward to measure this quantity, one usually simply measures the thickness of a "standard" sample. In the past, highly reflective Ag surfaces have been used as a reference for the absolute reflectance. A more reproducible surface, however, is a clean Si surface. The presence of thin native oxides on Si do not significantly affect the reflectance except in the UV.

For wavelength quality assurance, one can obtain several narrow-band transmission filters at various wavelengths and simply measure the transmittance through the filters (if the instrument can be configured to measure transmittance). Before using a filter as a standard, however, some method should be used to certify the wavelength of maximum transmittance. The filter manufacturer can often provide such certification for a slight additional cost. This service can also be obtained from NIST. A second method (in the UV region) is to use the critical features of a clean silicon wafer. A third method, which works over the entire spectral range, is to use the characteristic lines in the emission spectrum of the light source. The second and third methods do not require any additional standards, but do not have as high a resolution as the filter method.

The angle of incidence can be measured with spectroscopic ellipsometers very accurately at one particular angle. Figure 10.1 shows this affect. In this figure, we show the delta plot for a silicon substrate with several different thicknesses of native oxide. Note that although the overall curves are significantly different, the point where the curves cross 90° is somewhat insensitive to thin native oxides. Accordingly, a freshly cleaned silicon wafer (H-termination is not required) can be measured and the angle of incidence can be one of the regression parameters along with the thickness of the native oxide. The use of the angle of incidence as a regression parameter is usually reserved for the quality assurance measurements and is not normally done with ordinary analysis.

The angle of incidence for reflectometers is normally not an issue. The light is usually focused with a lens and is near-normal for all of the rays; hence small variations are insignificant.

There is one final recommended measurement which is not a quality assurance measurement, but is useful in the day-to-day operation of these measurement tools. As the light source ages, the intensity will gradually decrease. This will not affect the values obtained as long as there is sufficient light to obtain an adequate signal-to-noise ratio. In most cases, the intensity is smallest in the UV region, and it is this region where the degradation will be first noticed. If the

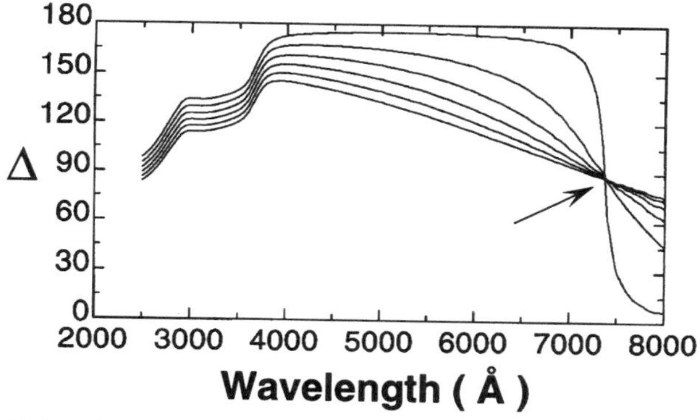

Figure 10.1 Delta plotted for a silicon wafer with native oxide with thicknesses of zero, 10, 20, 30, 40, and 50 Å. Note that the wavelength location where delta crosses 90° is reasonably insensitive to native oxides.

spectral output of the source is recorded monthly or quarterly, one can predict to some extent when a new source will be needed. For an instrument which can be used to measure transmission, this can readily be done by simply taking a baseline spectrum (the slit width and any aperture setting should be the same each time).

The question sometimes arises as to whether the light source should be turned off when the instrument will be idle for a period of time. Several tens of minutes or a few hours are needed for an arc source to stabilize. Most practitioners simply leave the source on all the time, even if the instrument is to be used only once a week.

10.5 REFERENCES

1. G. A. Candela, D. Chandler-Horowitz, J. F. Marchiando, D. B. Novotny, B. J. Belzer, and M. C. Croarkin, *Standard Reference Materials: Preparation and Certification of SRM-2530, Ellipsometric Parameters Δ and Ψ and Derived Thickness and Refractive Index of a Silicon Dioxide Layer on Silicon,* NIST Special Publication 260-109, US Government Printing Office, Washington, 1988.
2. T. Yasuda and D. E. Aspnes, *Appl. Opt.*, **33**, 7435 (1994).
3. P. Dumas and Y. J. Chabal, *Chem. Phys. Letters*, **181**, 537 (1991); S. Adachi and K. Utani, *Jpn. J. Appl. Phys.*, **32**, L1189 (1993); T. Suzuki and S. Adachi, *Jpn. J. Appl. Phys.*, **33**, 5599 (1994); S. Adachi, T. Ikegami, and K. Utani, *Jpn. J. Appl. Phys.*, **32**, 4398 (1993); Y. Morita and H. Tokumoto, *Appl. Phys. Lett.*, **67**, 2654 (1995); H. Yao, J. A. Woollam, and S. A. Alterovitz, *Appl. Phys. Lett.*, **62**, 3324 (1993).

CHAPTER 11

Very Thin Films

11.1 INTRODUCTION

There are some special considerations for very thin films. We shall consider films which are 100Å or less to be in this category. Some examples in this category are native oxides, some gate oxides on microelectronic devices, metal layers in magnetic random access memory devices, self-assembled monolayers, adsorbed layers, and so on.

The areas of consideration are:

1. Determining the thickness of a film where the optical constants are known.
2. Determining the optical constants of a very thin material.
3. Distinguishing one film material from another, either as a single layer or in a stack.

We shall find that ellipsometry, and in some cases reflectometry, deals with situation 1 very well. In most cases, situation 2 will be very difficult if not impossible. In some cases situation 3 can be handled well, and in some cases it is very difficult.

11.2 DETERMINING THICKNESS

11.2.1 For Reflectance

Reflectivity is relatively unaffected by the presence of a thin dielectric film. Figure 11.1 shows the reflectance plots for film-free single-crystal silicon and plots for silicon with 50 Å oxide and 100 Å oxide. Although some benefit can be gained by working in the UV region, for the most part, the reflectance technique is inappropriate for determining the thickness of very thin dielectric films.

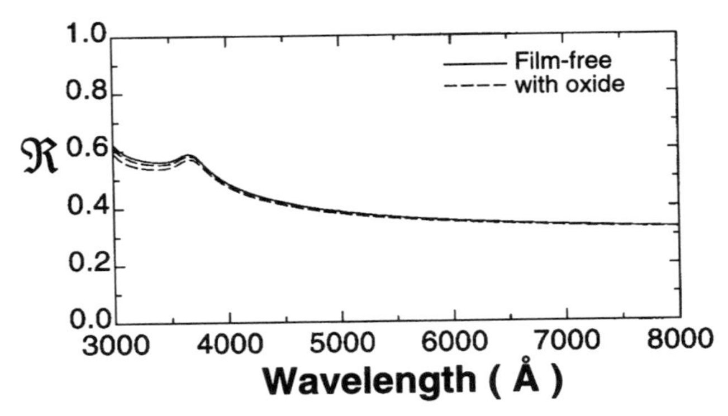

Figure 11.1 Reflectance from film-free silicon and for oxide on silicon. Plots for 50-Å and 100-Å oxide films are shown.

This can be used to advantage, however. Reflectance is the ratio of the intensity of the light being reflected from a surface to the light incident on that surface. It is often difficult to measure the incident light directly. Accordingly, the incident light is usually measured by reflecting the light from a reference. The presence of a thin native oxide on silicon will not affect the reflectance significantly, and hence a silicon wafer makes a very robust reference for determining the intensity of the incident light.

The presence of thin metal layers will make significant changes in the reflectance of a substrate. Figure 11.2 shows the reflectance from film-free silicon along with the reflectance from a thin chromium[1] layer. Plots for 25, 50, 70, and 100 Å chromium are shown.

Whereas for a thick dielectric film, the *shape* of the reflectance spectrum is significant (the spectral location of maxima and minima), this is not the case

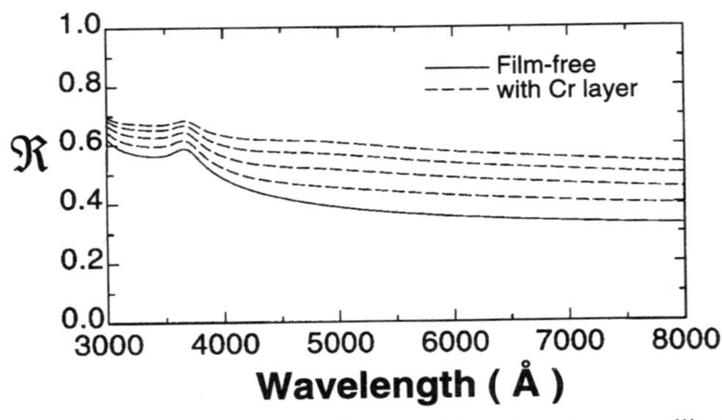

Figure 11.2 Reflectance for film-free silicon and for chromium on silicon. Plot for 25, 50, 75, and 100 Å of chromium are shown. Optical constants for chromium are taken from Palik.[1]

for thin metal films. As the thickness of a metal layer increases, the film will gradually become opaque and the reflectance will gradually change from that of the original substrate to that of a substrate of the metal. The constructive and destructive interference phenomena do not usually play a significant role; hence the shape of the curve as a function of wavelength is less important than simply the magnitude of the reflectance. Focusing errors in production tools will translate into errors in the thickness value obtained. This accentuates the need for good reflectance reference data and proper optical constants for the film material.

11.2.2 For Ellipsometry

When the optical constants are available, ellipsometry can readily be used to determine the thickness of very thin films. Figure 11.3 shows the delta spectrum for film-free silicon along with the delta spectra for several thin oxide films on silicon. We do not show the psi spectrum because there is very little change in psi for the first 100 Å of oxide growth.

The reason that ellipsometry is so sensitive to thin films is that, whereas the intensity of the reflected light is not affected very much by the presence of a thin dielectric layer, the phase difference between the s-wave and the p-wave is significant and measurable. Roughly speaking, psi carries the reflected intensity information, and hence is not affected much by thin films. Delta carries the phase information, and a thin film will induce a measurable change in delta.

A reasonable question might be how thin a film can be measured. With a film with index 2.0 on a silicon wafer, at an angle of incidence of 70° and a wavelength of 6328 Å, a change in delta of 0.25° represents a thickness change of about 1 Å. Since our instruments can routinely measure to an uncertainty better than this, one would expect that we should be able to measure monolayers.

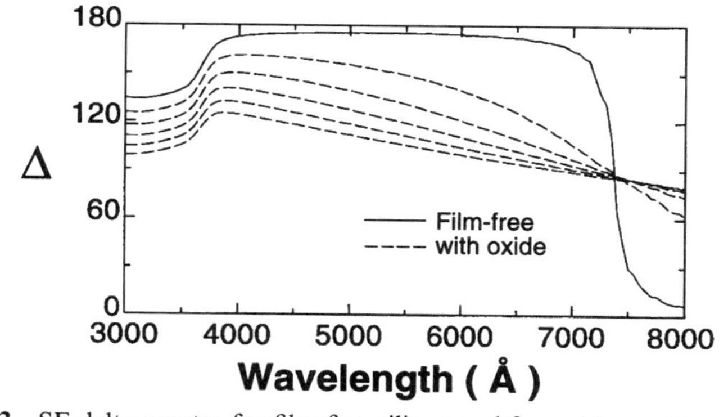

Figure 11.3 SE delta spectra for film-free silicon and for oxide on silicon. Plots for 20, 40, 65, 80, and 100 Å of oxide are shown. The angle of incidence is 75°.

In some of the earliest ellipsometric measurements, Drude[2] observed films on polished metal surfaces which were only a few angstroms thick. For very thin films of this nature, the Drude approximation[3] for delta is often used. It simply states that the change in delta is proportional to the thickness of the film, and is expressed as

$$\Delta^0 - \Delta = C_\Delta x \tag{11.1}$$

where Δ^0 is the film-free value for delta, x is the film thickness, and C_Δ is the proportionality constant. A more complete discussion of the Drude approximation is given by Tompkins.[4]

For films which are only a few angstroms thick, the change in Δ is more important than the absolute values of either Δ or Δ^0. These types of experiment are often done in situ, measuring Δ^0, adding the film without moving the substrate, and measuring the new value of Δ. It is seldom possible to measure the substrate, remove the sample, add the film, and return the sample to the ellipsometer. Spot-to-spot variations in a substrate or slight changes in angle of incidence may induce uncertainty in the value of Δ^0 which is greater than the change due to the film.

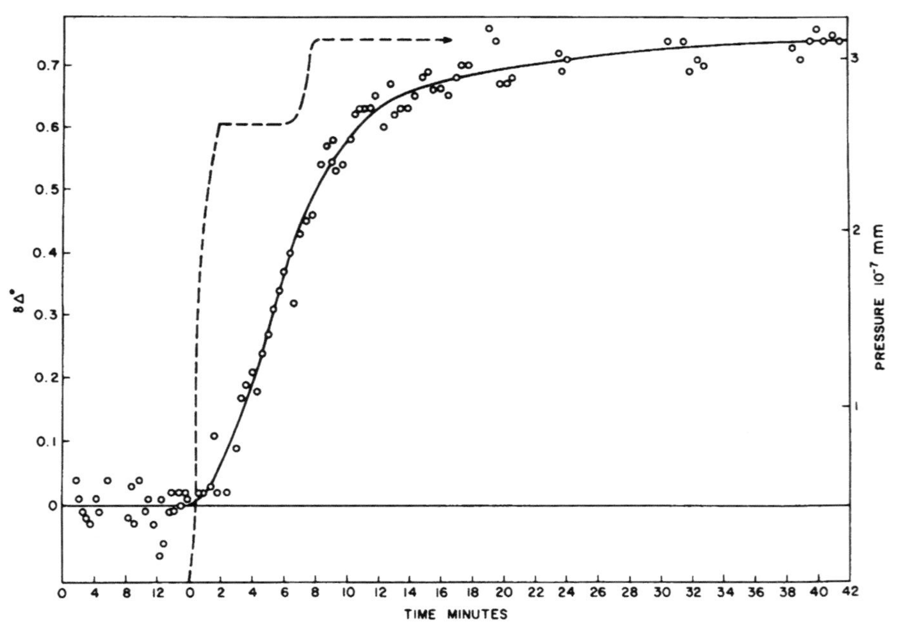

Figure 11.4 The change in Δ during chemisorption of oxygen on cleaved silicon under high vacuum at room temperature versus exposure time. After an initial steep rise, $\delta\Delta$ saturates when a complete monolayer has been formed. The dashed curve shows the rise in pressure as O_2 was suddenly admitted (after Archer[5]).

In one of the classical ellipsometry experiments, Archer[5] cleaved silicon under ultrahigh vacuum and allowed oxygen to chemisorb on the surface. He used ellipsometry to follow the progress of the chemisorption. The plot of the change in Δ versus exposure time is shown in Figure 11.4. For the exposure time prior to zero, we can see the scatter in the experimental measurements. At time zero, oxygen is admitted, the pressure rises to about 2.7×10^{-7} Torr, and the chemisorption begins. At time 8 min, more oxygen is admitted, raising the pressure to slightly over 3×10^{-7} Torr. The change in Δ saturates at about 0.7°, indicating a complete monolayer. Note that in this experiment, there is enough sensitivity to measure fractions of a monolayer coverage.

Whereas dielectric films cause primarily a change in the phase difference, a thin *metal* film will cause a significant change in intensity. Accordingly, we would expect that a thin metal film would produce a change in both delta and psi, and this is the case.

11.3 DETERMINING OPTICAL CONSTANTS

The basis for determining optical constants of dielectric films in single-wavelength ellipsometry (SWE) is illustrated in Figure 11.5 (also shown as Figure

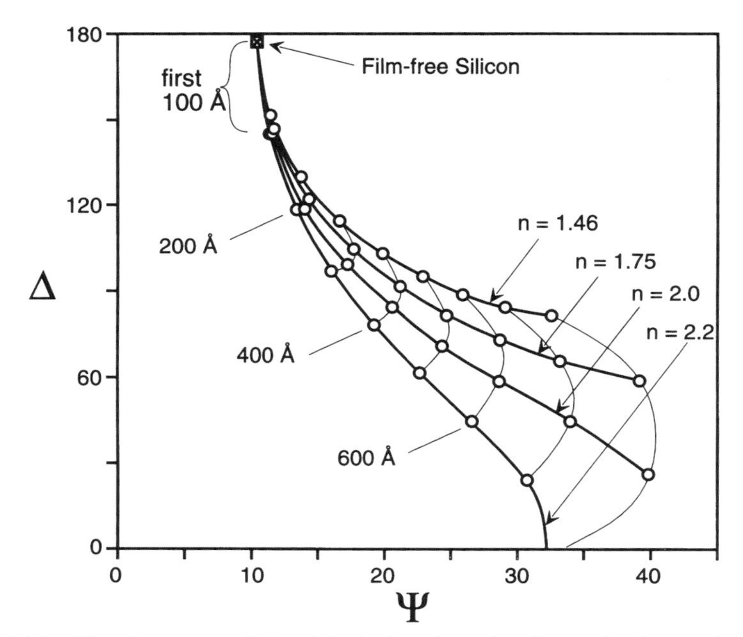

Figure 11.5 The first part of the delta/psi trajectories for a single wavelength for transparent films on silicon with index of refraction of the films as indicated. These trajectories are for an angle of incidence of 70° and a wavelength of 6328 Å.

6.7). As the delta/psi trajectory moves away from the substrate point, the trajectories diverge, depending on the optical constants. When the thickness is greater than a few hundred angstroms, the trajectories are well separated and it is easy to determine the index by observing which trajectory the measured delta/psi point falls on. Note, however, that for films which are less than 100 Å thick, the curves are very close together and it is difficult, if not impossible, to distinguish one curve from another. In addition, small errors in the value of psi will cause large errors in the index value. In this regime, changes in delta can be used to determine the product of index × thickness (in Eq. 2.19, β is a function of the product of thickness and index). If the index is known, the thickness can readily be determined, as discussed in the last section.

Accordingly, in SWE it is unwise to attempt to determine the index of refraction of a dielectric film which is very thin. Some SWE manufacturers recommend that the index is not determined for dielectric films less than 500 Å thick.

For spectroscopic ellipsometry the situation is somewhat different, although some of the same difficulties are experienced. For a given substrate, there is a substrate-point and a set of trajectories for each wavelength used and they are not necessarily the same. As is usually the case, working with shorter wavelengths (e.g., in the UV) allows us to measure thinner films before difficulties begin to arise.

There are some other advantages of SE over SWE. In a typical SWE measurement, one measures one value of delta and one value of psi and hopes to calculate one value of thickness and one value of the index of refraction. For SE, one typically measures 100 values of delta and 100 values of psi. For a point-by-point analysis (i.e., the various values of the index are not considered related to each other) one hopes to determine 100 values of n, but only one value of thickness. When using a dispersion relationship (i.e., a Cauchy equation), the number of values to be determined is still further reduced, and the effect of some of the random errors may be averaged out.

The fact remains, however, that the trajectories for various indices are quite close together for very thin films. Accordingly, we do not recommend attempting to determine optical constants for films which are less than 100 Å thick.

Instead, in order to determine thicknesses, one should choose optical constants of a material similar to the material of interest. One must understand that thicknesses determined in this manner are *equivalent thicknesses* commensurate with this assumption.

11.4 DISTINGUISHING BETWEEN MATERIALS

As an example of distinguishing one material from another, let us consider a gate oxide film on silicon which, when modeled with index of refraction values in tabular form, was determined to be 142 Å thick. The measured and modeled

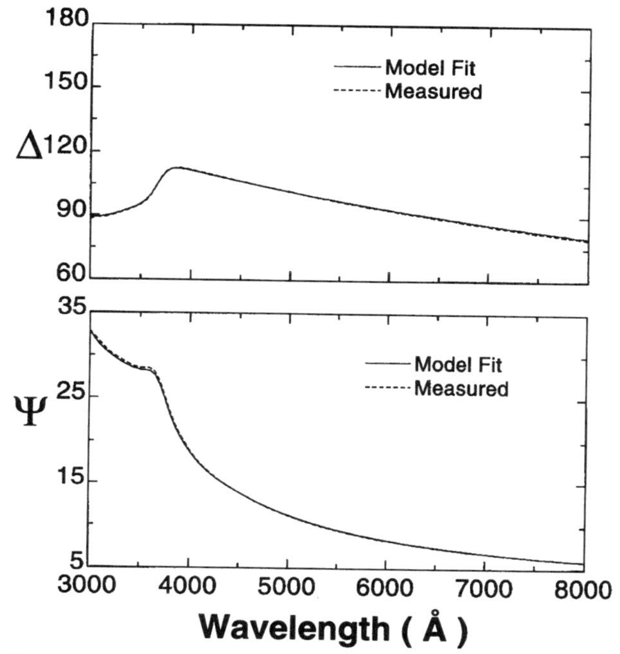

Figure 11.6 A gate oxide on silicon, modeled as SiO_2, with tabular values for the index of refraction.[6] The thickness value was determined to be 142 Å with an MSE of 5.4.

values are shown in Figure 11.6. In this case, the MSE is 5.4, implying a very good fit.

Suppose that this were an unknown film and we modeled it as a nitride film using literature index values. The measured and modeled values are shown in Figure 11.7. The thickness value obtained is 112 Å and the MSE is 61. Although for SWE it would be difficult to tell whether a film of this thickness was an oxide or a nitride because of the experimental uncertainty of the measurement, for SE we have significantly more data, and the experimental uncertainty of a single wavelength measurement is not as limiting. In this case, we could clearly see that the nitride model is inappropriate and the oxide model fits nicely.

On the other hand, let us consider a sample which is much thinner (a native oxide on silicon). When modeled as an oxide, using the tabular index of refraction data, it is determined to be 25.5 Å thick, with an MSE of 6.4. The measured and modeled values are shown in Figure 11.8.

Let us consider the situation if this were an unknown film. Whereas the 142 Å oxide could readily be distinguished from a nitride film, in this case with the very thin film, there are alternative models all of which give reasonable fits. Table 11.1 gives the thickness and MSE for four films, all of which fit the data

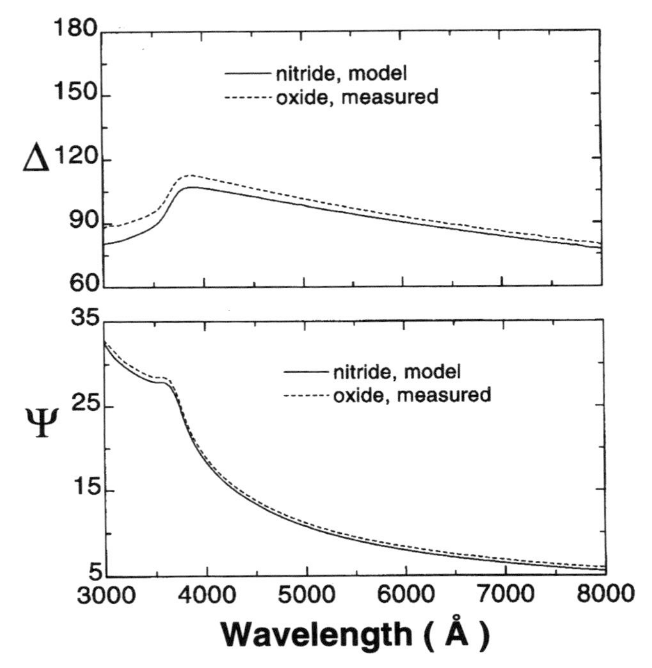

Figure 11.7 The sample used in the previous figure, treated as an unknown film, modeled as Si_3N_4, with tabular values for the index of refraction.[6] With this assumption, the thickness was determined to be 112 Å with a MSE of 61.

well. The *roughness* layer was modeled as half silicon and half air (see the chapter on roughness). Figure 11.9 shows the modeled curves for all four of these possibilities. The overlap is such that it is difficult to tell that there are four curves on the plot.

The point illustrated here is that for films which are thicker than 100 Å, many films can be distinguished from each other. For very thin films, there are often several models which give similar fits, so it is not possible to distinguish one material from another.

In order to distinguish one material from another, it is necessary that they have optical constants which are significantly different. Accordingly it is very easy to distinguish a thin metal film from a thin dielectric film because of the different shape of the dispersion curve. On the other hand, most dielectrics have dispersion curves which have similar shapes (with different magnitudes). Accordingly, it is very difficult to distinguish a very thin film of one dielectric from a very thin film of another dielectric.

Another ramification of this insensitivity for thin films is that if one has a stack of two or more very thin films which have dispersion relationships which have a similar shape, it is very difficult to determine the thickness of each layer

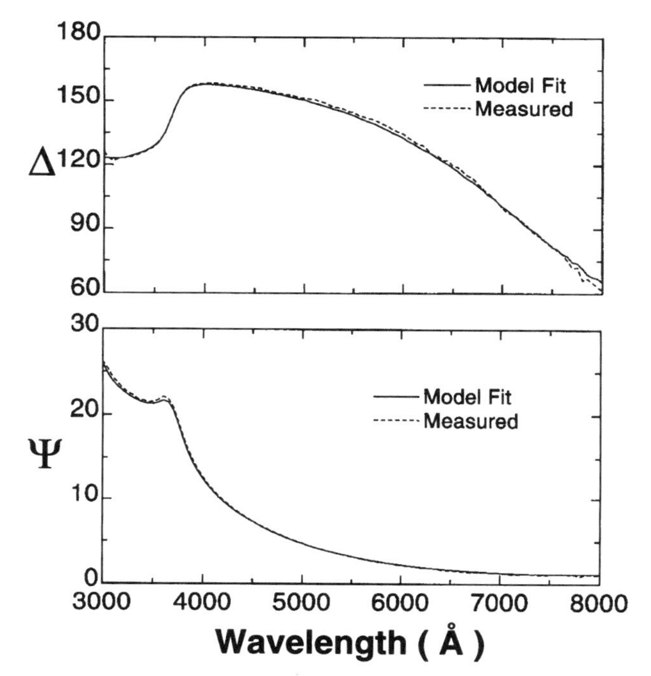

Figure 11.8 A native oxide film on silicon, modeled as SiO_2, with tabular values for the index of refraction.[6] The thickness value was determined to be 25.5 Å with an MSE of 6.4.

unambiguously. An example of this might be a thin layer of aluminum oxide on a thin layer of silicon nitride on a silicon wafer. On the other hand, if the materials have dispersion relationships which are very different, one can readily determine the individual thicknesses. Examples of this situation might be a thin film of aluminum on silicon covered by a thin film of aluminum oxide.

TABLE 11.1 Four Materials which Give a Reasonable Fit to the Measured Data Shown in Figure 11.8

Material	Thickness (Å)	MSE
Oxide	25.5	6.4
Nitride	20.1	8.4
Roughness	21.6	12.1
Polyethylene	24.9	6.5

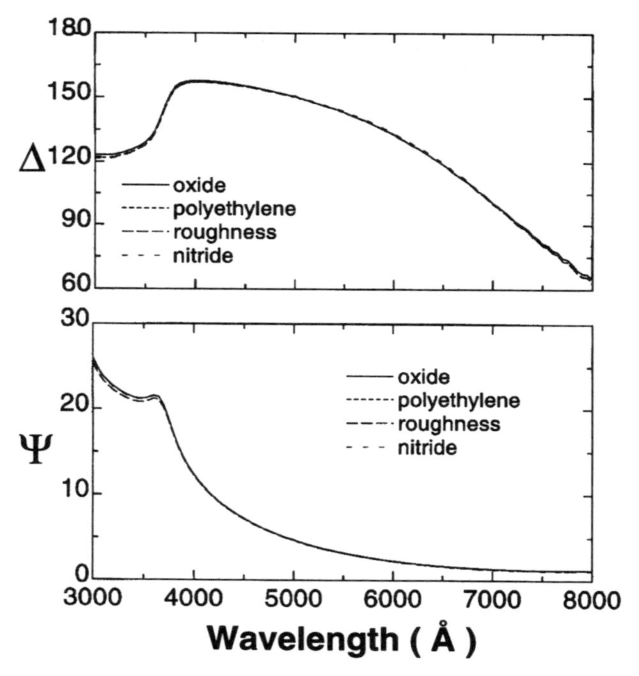

Figure 11.9 Four different models which give reasonable fits to the data shown in the previous figure.

11.5 REFERENCES

1. E. D. Palik (editor), *Handbook of Optical Constants of Solids II*, Academic Press, New York, 1991.
2. P. Drude, *Ann. Phys. Chem.*, **36**, 865 (1889).
3. A. N. Saxena, *J. Opt. Soc. Am.*, **55**, 1061 (1965).
4. H. G. Tompkins, *A User's Guide to Ellipsometry*, Academic Press, New York, 1993.
5. R. J. Archer, in *Ellipsometry in the Measurement of Surfaces and Thin Films*, edited by E. Passaglia, R. R. Stromberg, and J. Kruger, National Bureau of Standards Miscellaneous Publication 256, US Government Printing Office, Washington, 1964, pp. 255 ff.
6. E. D. Palik (editor), *Handbook of Optical Constants of Solids*, Academic Press, New York, 1985.

CHAPTER 12

Roughness

12.1 INTRODUCTION

In this chapter we consider the effects of roughness on ellipsometric and reflectance measurements. This roughness may be at the surface of the sample and/or at the interfaces between layers in the sample. It is important to note that it is usually not our intent to characterize the roughness of the sample under study; rather we wish to ensure that the presence of roughness does not negatively impact the accuracy and precision of film thickness and optical constant measurements.

12.2 ROUGHNESS IN GENERAL

There are two ways in which surface or interfacial roughness affects optical measurements. The first effect is somewhat subtle. Consider a rough surface as shown in Figure 12.1.

The bulk material (below the bottom dashed line) has well-defined optical constants. The ambient (above the top dashed line) also has well-defined optical constants. The region between the two dashed lines may or may not have well-defined optical constants, and if it does, they will certainly not be the same as those of the bulk material or the ambient. We can consider the region bounded by the dashed lines to be (effectively) a film with well-defined optical constants which are different from the bulk material and the ambient. This approximation is valid if the width of the region (between the dashed lines) is much less than the wavelength of light used for the measurement.

In this case, the presence of roughness at the surface has a similar effect to the presence of a thin film at the surface, where the thin film has optical constants somewhere in between those of the regions above and below the interface. For this reason it is adequate in many cases to model surface or interfacial roughness as a discrete thin film at the surface or interface. It is particularly important to model the surface roughness in this manner when determining the optical constants of substrates or films by ellipsometry.[1]

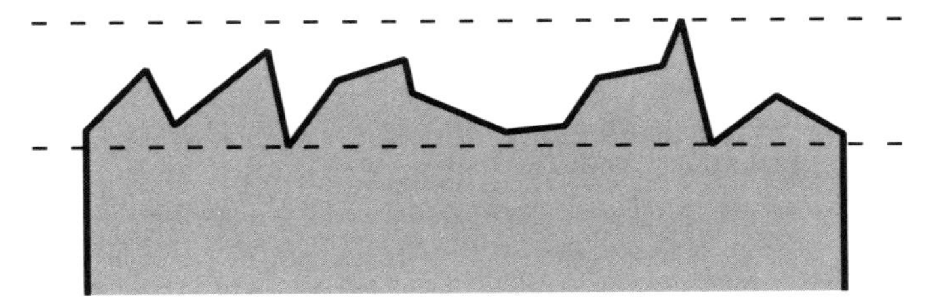

Figure 12.1 Rough surface, shown schematically.

The second effect caused by roughness is nonspecular scattering of the incident light. As can be expected, this is much more of a problem for reflectometry than ellipsometry. Reflectometry measures the ratio of the intensity of the reflected beam to the incident beam, so scattering of light away from the detector will affect the measurement of this ratio. This scattering loss will tend to be larger for rougher surfaces, and will tend to be larger for shorter wavelengths than for longer wavelengths. The scattering loss will then lead to a reduction in the measured reflectance, and this reduction will be wavelength-dependant.

Scattering loss has much less of an effect on ellipsometric measurements owing to the fact that ellipsometers measure the ratio of the p- and s-complex reflectances of the sample, rather than the absolute reflectance of the sample. Problems can occur, however, if the sample is rough enough to cause depolarization of the reflected light beam. Recall that for an isotropic sample, energy cannot be transferred by the sample from the p- to the s-polarization state, and vice versa. Very rough samples can cause the transfer of energy between the polarization states, thus depolarizing the reflected beam. This effect can be very difficult to model, and will tend to be strongly dependant on the topography of the rough surface, the materials involved, the angle of incidence, and the wavelength. For this reason ellipsometry is generally not considered to be very effective for very rough samples.

12.3 APPROXIMATION OF SURFACE ROUGHNESS AS AN OXIDE LAYER

In the case where there is not a lot of roughness on the surface (< 100 Å or so), it is often possible to model the roughness fairly well as an oxide film. This is particularly useful for polycrystalline and amorphous silicon films and metal films, where some degree of surface oxidation will exist in addition to the surface roughness. It is nearly impossible to separate the effects of the roughness from the presence of the thin oxide at the surface. This is beneficial, as usually we do not actually wish to measure the roughness. Rather we want to model the surface effects properly in order to get accurate measurements of what lies underneath the surface.

Two approaches are possible when using thin oxide layers to model surface roughness/oxidation. In many cases the measured data will be sufficiently sensitive to the thickness of the surface layer to allow the surface oxide layer thickness to be varied without causing correlation problems in the model. This will usually not be the case if there exists a high degree of variability (particularly for optical constants) in the rest of the optical model. If the thickness of the surface oxide cannot be varied due to correlation issues, we generally fix the oxide thickness at a physically reasonable value (usually 20–30 Å).

12.4 EFFECTIVE MEDIUM APPROXIMATIONS (EMAs)

As mentioned previously, surface or interfacial roughness can be thought of as a layer consisting of a mixture of the two materials above and below the interface (one of these materials is air when considering surface roughness). There is a broad class of dispersion models, known as effective medium approximations (EMA), which can be used to predict the optical constants of mixtures of materials of known optical constants. The simplest EMA

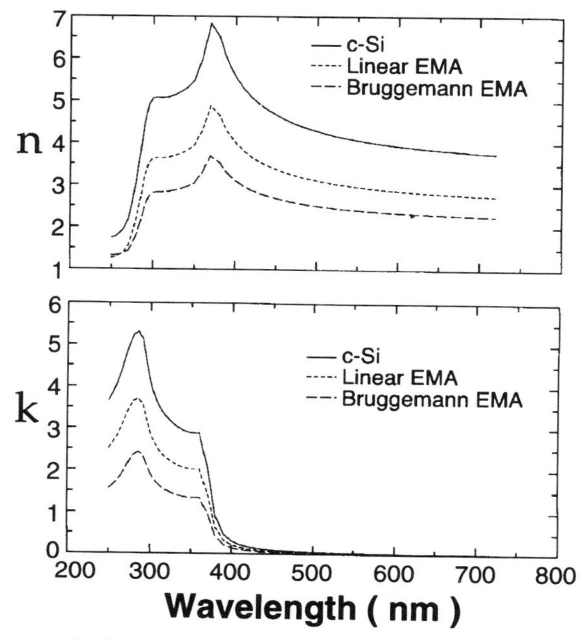

Figure 12.2 Index of refraction n and extinction coefficient k of c-Si and of a mixture of 50% c-Si and 50% voids. Spectra were calculated using both linear and Bruggemann EMA models.

model consists of linearly interpolating between the optical constants of the constituents:

$$\tilde{\varepsilon} = f_A \tilde{\varepsilon}_A + f_B \tilde{\varepsilon}_B \tag{12.1}$$

where $\tilde{\varepsilon}$ is the effective complex dielectric function of the composite material, $\tilde{\varepsilon}_A$ and $\tilde{\varepsilon}_B$ are the complex dielectric functions of the constituent materials, and f_A and f_B are the volume fractions of the two constituent materials. Note that $f_A + f_B$ must equal unity. The linear EMA, while simple to calculate, is not particularly accurate. The most commonly used EMA model is the Bruggeman EMA, in which the effective dielectric function of the composite material is obtained by solving the following equation:

$$f_A \frac{\tilde{\varepsilon}_A - \tilde{\varepsilon}}{\tilde{\varepsilon}_A + 2\tilde{\varepsilon}} + f_B \frac{\tilde{\varepsilon}_B - \tilde{\varepsilon}}{\tilde{\varepsilon}_B + 2\tilde{\varepsilon}} = 0 \tag{12.2}$$

Figure 12.2 shows the results of using the linear and Bruggemann EMA models to calculate the optical constants of a mixture of 50% crystalline silicon and 50% voids.

Generally, unless the roughness on the sample under study is very thick (> 100 Å, for example), the choice of EMA model is not critical to the model.

12.5 EXAMPLES

12.5.1 Polysilicon

A common example of analysis requiring the modeling of surface roughness is polysilicon. Figure 12.3 shows the fit to measured absolute reflectance data from a 2000 Å thick polysilicon film deposited on 1000 Å of thermal silicon

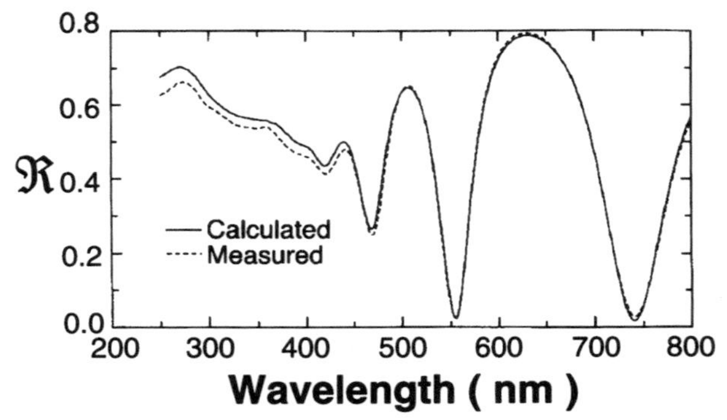

Figure 12.3 Measured and best-fit calculated absolute reflectance from a 2000-Å polysilicon film deposited on thermal oxide on silicon. Surface roughness was not modeled.

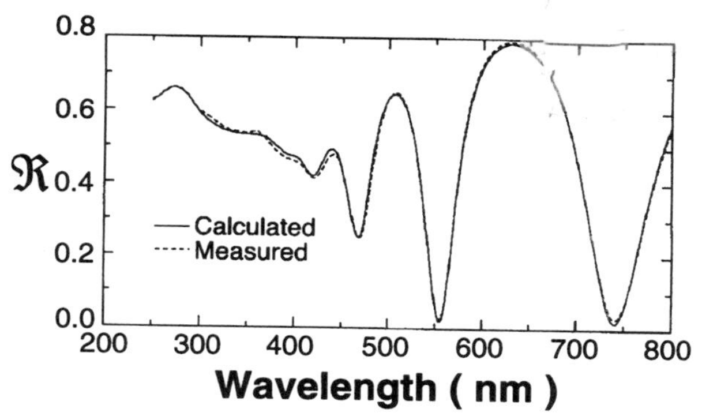

Figure 12.4 Same measured data as in Figure 12.3. Best model calculated using surface roughness layer.

dioxide on a silicon wafer. Note that surface roughness was not modeled in this analysis.

Note that the data fit quite well in the visible, but the fit is rather poor at the shorter wavelengths. Figure 12.4 shows the same measured data, this time fit using the same model but with the inclusion of a surface roughness layer.

12.5.2 Sensitivity in the Visible

In contrast to this case, ellipsometric data can be very sensitive to roughness in the visible part of the spectrum. This is because the presence of roughness can strongly effect the *shape* of the measured delta spectrum. Figure 12.5 shows the

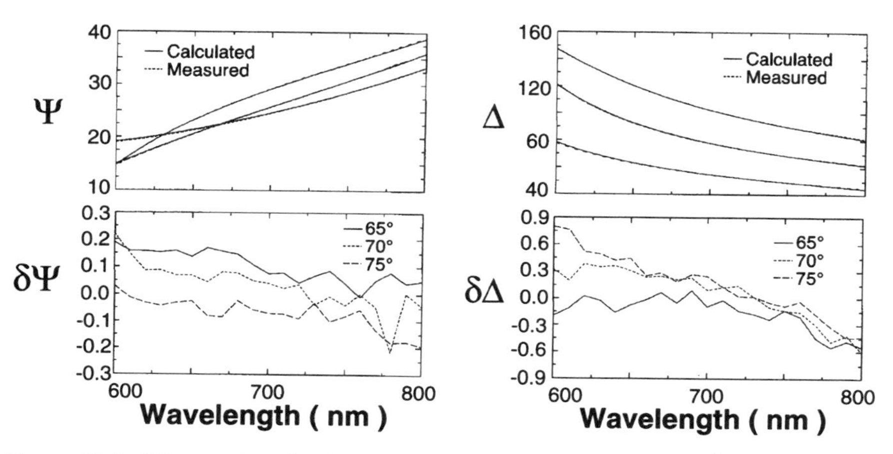

Figure 12.5 Measured and calculated Ψ and Δ data for an 1800 Å organic film on silicon. The difference between the measured and calculated data is shown in the lower graphs as $\delta\Psi$ and $\delta\Delta$.

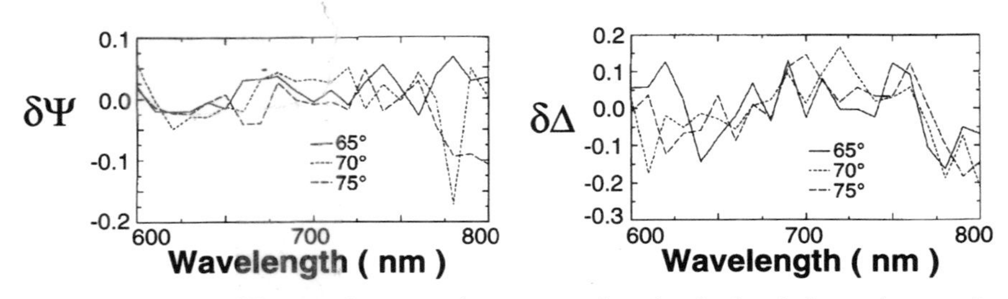

Figure 12.6 The difference between the measured and calculated data when surface roughness is included. The measured data are the same as in Figure 12.5, but surface roughness was included in the optical model.

measured and best-fit calculated psi and delta data for an organic film modeled as a transparent layer, without surface roughness. To illustrate small differences, we also show the difference between the measured and modeled data as $\delta\Delta$ and $\delta\Psi$.

If we now include a surface roughness layer in the model, using the Bruggemann EMA, the fit is improved. This is shown in Figure 12.6, where the roughness layer is fit to 51.9 Å.

Note that a considerably better fit was obtained using the surface roughness in the model, as illustrated in the difference plots. Note the difference in vertical scale between the difference plots with and without roughness in the model, and that the difference when roughness is included is not systematic.

12.5.3 Roughness Effects when Determining Optical Constants

The proper inclusion of roughness in the model is particularly important when determining optical constants from ellipsometric data. Using the thicknesses determined in the previous example, we do a point-by-point fit of the data over the entire spectral range in order to determine the optical constants. Figure 12.7 shows a comparison of optical constants obtained for the organic film when including the roughness layer and when ignoring the roughness layer. Obviously, it is very important to model surface roughness properly in order to obtain accurate optical constants from thin films.

12.5.4 SOI Interface Roughness

The final example illustrates modeling of both surface and interfacial roughness. The sample under study in this case is an implanted silicon-on-insulator (SOI) structure, consisting of an epitaxial silicon layer on oxide on silicon. Owing to the means by which this structure was produced, there is a significant interfacial layer between the silicon substrate and the oxide layer which must be included in the optical model. The structure is shown in Figure 12.8. Figure 12.9 shows the results of the analysis of absolute reflectance data from this

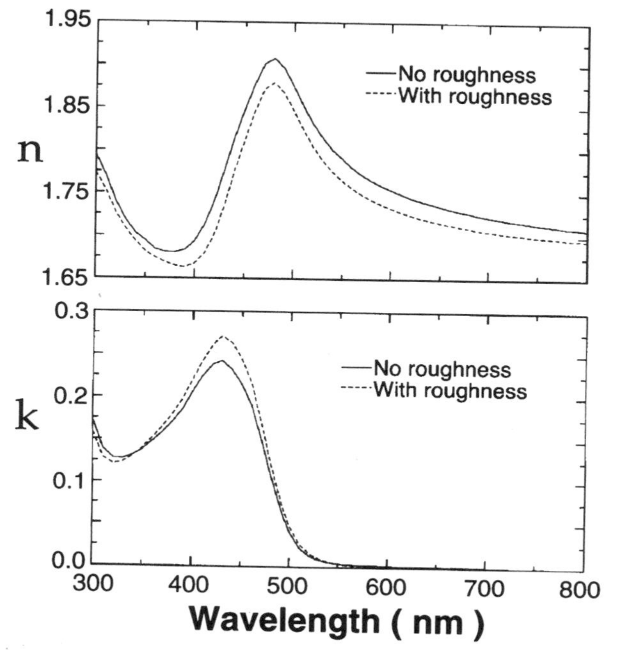

Figure 12.7 Optical constants obtained for the organic film with and without surface roughness included in the optical model.

structure when roughness is not included in the model. This analysis yields a top silicon layer thickness of 3433 Å and an oxide thickness of 3912 Å. There clearly is some serious misfit between the measured and modeled data.

Let us now include an interface layer with 50% silicon and 50% oxide. This introduces one additional regression parameter, the thickness of the roughness layer. The modeled and measured data are shown in Figure 12.10.

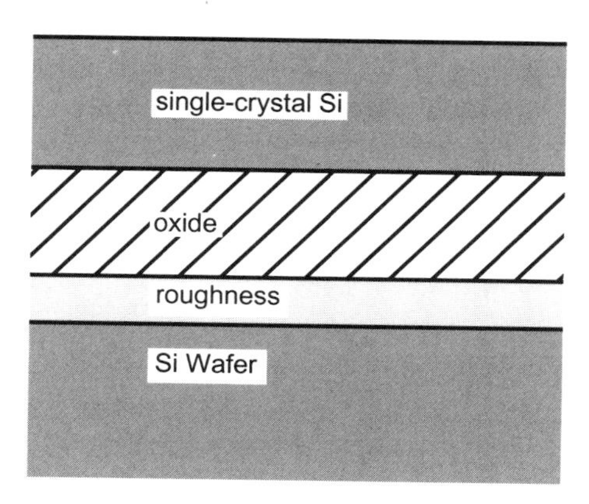

Figure 12.8 The silicon-on-insulator (SOI) structure, shown schematically.

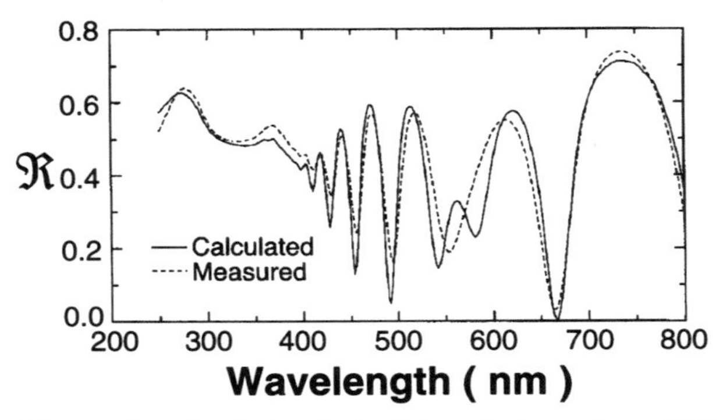

Figure 12.9 Measured and calculated absolute reflectance for a SOI structure. The interfacial region between the oxide and silicon substrate was not modeled.

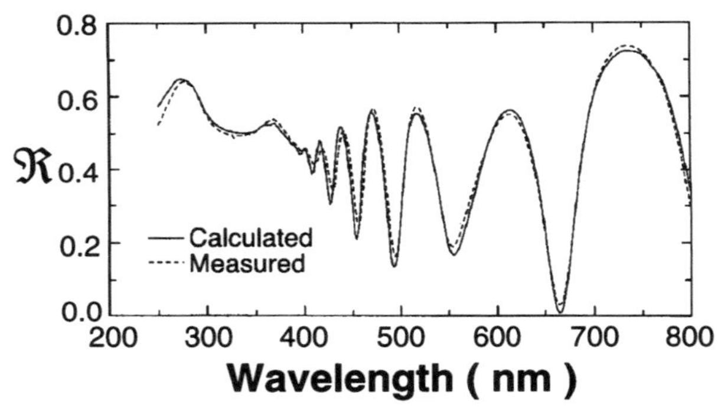

Figure 12.10 Measured and calculated absolute reflectance for a SOI structure. The interfacial region between the oxide and silicon substrate was determined to be 399 Å thick.

The thicknesses obtained when the interfacial region was modeled were 3427 Å for the epi-Si layer, 3652 Å for the oxide layer, and 399 Å for the roughness layer. Note that the thickness of the upper silicon layer is about the same for both models. It is evident from Figures 12.9 and 12.10 that including the roughness interface significantly improves the fit. The fit can be further improved slightly if we allow the fraction of silicon in the roughness layer to vary. The resulting interfacial layer then consists of 33.4% c-Si and 66.6% oxide.

12.6 REFERENCES

1. D. E. Aspnes, in *Handbook of Optical Constants of Solids*, edited by E. D. Palik, Academic Press, New York, 1985, p. 89.

PROTOTYPICAL ANALYSES

The following are examples of typical problems which can be dealt with using either ellipsometry or reflectometry or both. The intent is to suggest a step-by-step approach.

PROTOTYPICAL ANALYSIS No. 1

Thermal Oxide, LPCVD Nitride, or Photoresist on Silicon

PA1.1 SALIENT FEATURE

The salient feature of this prototypical analysis is that *the optical constants of these materials are available in tabular form*. The analysis consists of finding the correct thickness of the film only, and does not involve any determination of optical constants. This is the simplest of all analyses. We illustrate the process with silicon dioxide on silicon, although the procedure is the same for the other two examples mentioned in the title.

PA1.2 THINNER FILMS

The first example is an oxide where the thickness to be determined will turn out to be 223 Å. An ellipsometric analysis will consist of collecting the data (in this example, over a wavelength range of 3000–8000 Å at an angle of incidence of 75°) and proposing a seed thickness. The parameters used during processing would normally suggest a seed thickness which is approximately correct.

To exaggerate the analysis, let us suppose that we propose a seed thickness of 50 Å. The regression analysis calculates the model data for this thickness, compares it with the measured data, and computes a mean squared error (MSE). Figure PA1.1 shows the corresponding curves for delta and psi. The regression analysis tries different values of thickness, each time calculating the MSE, and finds that increasing the thickness from 50 Å reduces the MSE. The curve for 150 Å is also shown in the figure. Eventually, the regression analysis determines a thickness value of 223 Å, which gives a MSE of 4 and this is the minimum value. The calculated values are not shown in the figure since they virtually overlay the measured values.

Figure PA1.2 shows the reflectometry data (at normal incidence) for the same film along with the calculated data for the two postulated thicknesses mentioned above. There is very little change in the reflectance for the higher

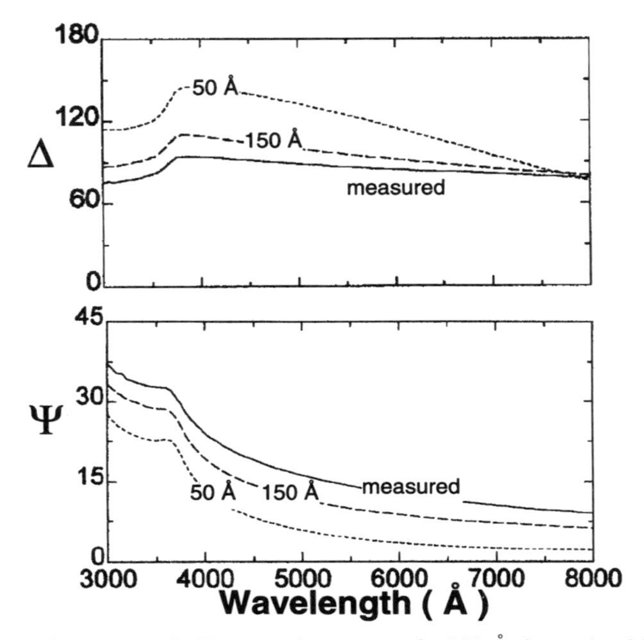

Figure PA1.1 The measured ellipsometric spectra of a 223-Å thermal oxide on silicon. Also shown are the calculated ellipsometric spectra for a 50-Å film and a 150-Å film.

wavelengths, but in the UV, there is sufficient difference for a thickness determination. The reflectometry instrument may use the same type of curve fitting procedures as the ellipsometry instrument, or may simply use a single wavelength for this determination. Since the shapes of the three curves are similar, the determination depends strongly on the intensity values. Slight intensity variations in the source will significantly affect the result; hence it is necessary to measure a reference sample frequently to determine the spectral intensity of

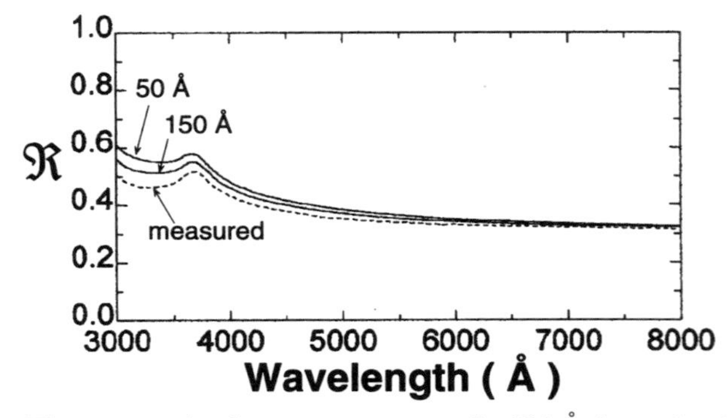

Figure PA1.2 The measured reflectance spectrum of a 223-Å thermal oxide on silicon. Also shown are the calcuated ellipsometric spectra for a 50-Å film and a 150-Å film.

the incoming light beam. Also, focusing errors in production tools will affect the measurement.

PA1.3 INTERMEDIATE FILMS

The second example is a film where the thickness to be determined will turn out to be 2830 Å. The measured data and calculated data for thicknesses of 2500 Å and 2700 Å are shown in Figure PA1.3. The analysis procedure is the same as before, that is, to obtain the measured data and to propose a seed value for thickness. In this case, if the proposed seed value is between 1445 and 3000 Å, the regression analysis will find the global minimum for the MSE at a thickness of 2830 Å, with a MSE of 13. If the seed value is outside this range, a local minimum rather than a global minimum will be found. The analyst can readily recognize this situation by observing that the MSE is much larger than normal or by simply observing that the measured and modeled curves do not fit very well.

Figure PA1.4 shows the reflectometry data for this film. Although there are only two minima and one maximum in the measured wavelength range, curve-fitting procedures can readily fit this curve and determine the thickness.

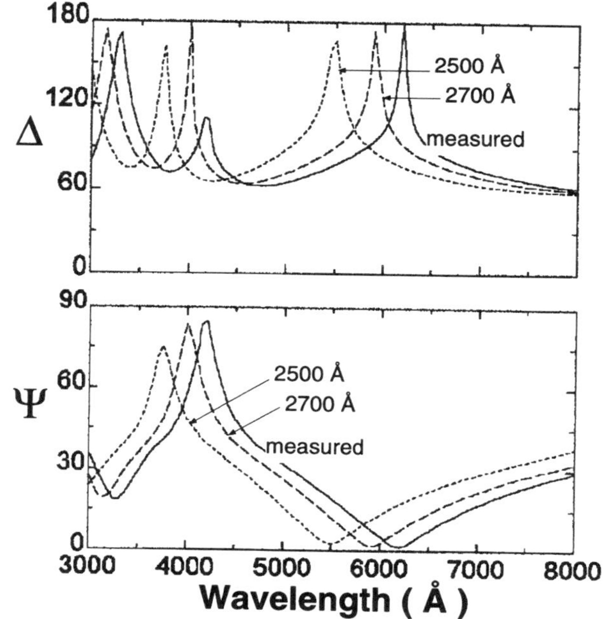

Figure PA1.3 The measured ellipsometric spectra of a 2830-Å thermal oxide on silicon. Also shown are the calculated ellipsometric spectra for a 2500-Å film and a 2700-Å film.

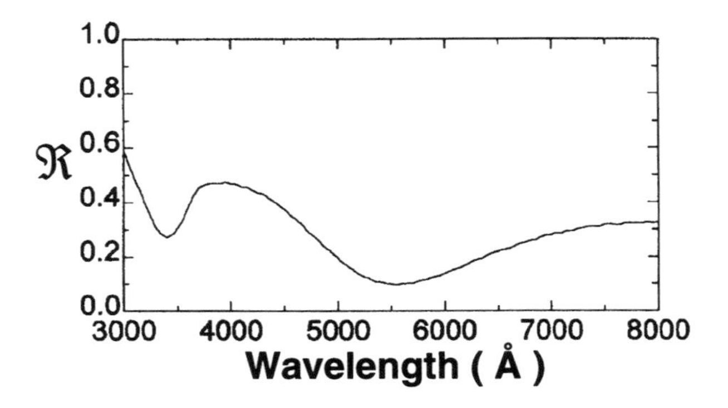

Figure PA1.4 The measured reflectance spectrum of a 2830-Å thermal oxide on silicon.

PA1.4 THICKER FILMS

The last example in this section is an oxide film which is somewhat thicker than the previous one. Ellipsometric measured data are shown in Figure PA1.5. The analysis is similar to the film described above. When the proposed seed value is between 6200 and 7780 Å, the regression analysis finds the global minimum for a thickness of 7397 Å. If a seed value outside this range is chosen, a local minimum will be found rather than a global minimum.

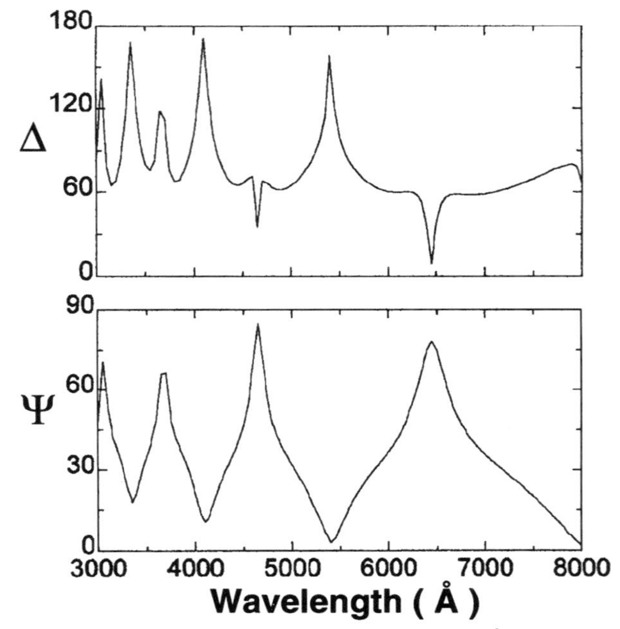

Figure PA1.5 The measured ellipsometric spectra of a 7397-Å thermal oxide on silicon.

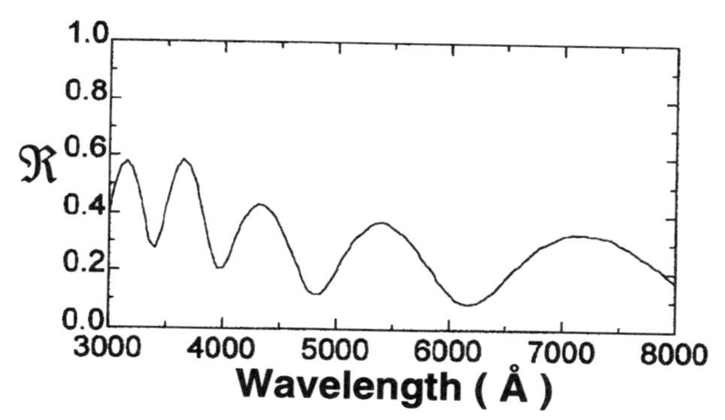

Figure PA1.6 The measured reflectance spectrum of a 7397-Å thermal oxide on silicon.

Figure PA1.6 shows the reflectance data for the same film. For this thicker film, there are several maxima and minima in the measured spectral range. The locations of the maxima and minima can be determined by the analytical software and the thickness determination can be made by these spectral locations. A further refinement is for the instrument to use the maximum/minimum spectral locations to determine the seed value for the full curve-fitting procedure.

PROTOTYPICAL ANALYSIS No. 2

Silicon Oxynitrides, PECVD Silicon Oxides, and Polysilicon

PA2.1 SALIENT FEATURE

The salient feature of this prototypical analysis is that *the optical constants of the material being characterized can be described as a mixture of the tabulated optical constants of two (or more) known materials.* An effective medium approximation[1,2] (EMA) is used to calculate the optical constants of the film from the known optical constants of the constituents and the percentage of each constituent. This type of analysis differs from those in the previous section in that the optical constants of the film under study can be adjusted in order to fit the measured data by varying the percentages of the constituent materials which are being "mixed." This type of analysis works best when the film actually consists of discrete regions of two or more materials, but can also work well when a material is expected to have stoichiometry or crystallinity in between two well-known endpoints, such as polysilicon. We illustrate this type of analysis with silicon oxynitride films (mixing silicon dioxide, silicon nitride, and perhaps silicon monoxide), PECVD silicon oxides (mixing silicon dioxide and silicon monoxide), and polysilicon (mixing crystalline silicon, amorphous silicon, and voids).

PA2.2 POLYSILICON

We begin with the analysis of polysilicon films, as they provide one of the simplest examples of the EMA type of analysis. To a reasonable degree of approximation, polysilicon can be considered to be a physical mixture of regions of crystalline silicon and regions of amorphous silicon. Grain boundaries and defects in the crystallites are considered to make up part of the amorphous silicon fraction. The degree to which this approximation is accurate depends on the microstructure of the polysilicon film and on the spectral range of the experimental data which are being analyzed. To illustrate this, Figure PA2.1 shows ellipsometric psi and delta data acquired from a polysilicon film created by depositing amorphous silicon onto a thin (56 nm) thermal oxide film, and then annealing the amorphous silicon film to produce polysilicon.

In this analysis, the parameters to be determined were the polysilicon and oxide thicknesses, as well as the percentage of amorphous silicon in the polysilicon film. In addition, it is invariably necessary to include a surface oxide or surface roughness on top of the polysilicon film as well. The thickness of the surface oxide was also a parameter to be determined. This analysis yielded thicknesses of 19.6 Å for the surface oxide, 631.1 Å for the polysilicon layer, and 561.3 Å for the thermal oxide layer. Also, the amorphous silicon fraction in the polysilicon layer was found to be 15.7%.

This type of analysis for polysilicon has the advantage of being very fast and robust, as the model for the optical constants used for the polysilicon layer has only a single variable parameter. It also has the added feature of producing a measured value (amorphous silicon percentage) which is indicative of the *quality*, or crystallinity, of the polysilicon layer. It is good to remember, however, that the EMA model, as the name implies, is an approximation (effective medium approximation), and there will be varying degrees of systematic error introduced into the results owing to the use of this approximation. In

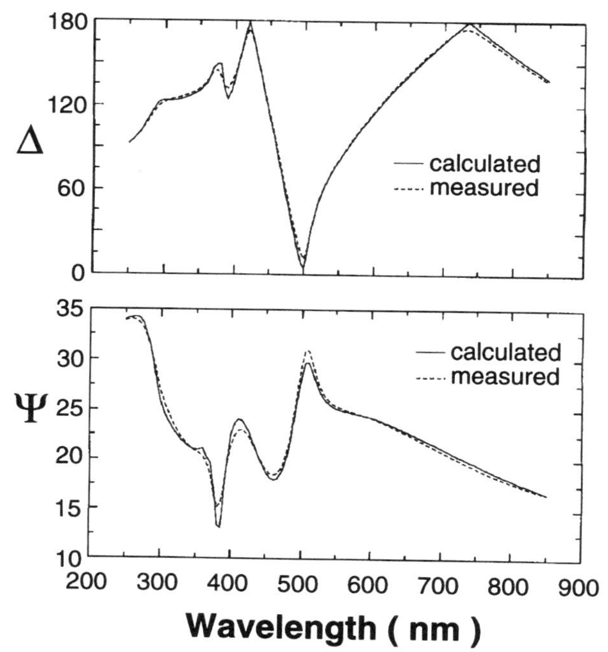

Figure PA2.1 Measured and calculated ellipsometric psi and delta data for a thin polysilicon film on thermal oxide on silicon.

production applications, this is not of primary concern, as the use of the EMA model will tend to yield the most reproducible measurements. To illuminate the types of error which may be encountered, Figure PA2.2 shows the same ellipsometric data as shown in Figure PA2.1. Here we have employed a very sophisticated dispersion model to describe the polysilicon optical constants, yielding thicknesses of 19.9 Å for the surface oxide, 630.0 Å for the polysilicon layer, and 536.2 Å for the bottom oxide.

Note that we achieve much better correspondence between the calculated and measured ellipsometric data. The dispersion model used in the second figure contained over 20 variable parameters, however, compared with the single variable parameter in the EMA model. The EMA model is not the ultimate model for accuracy, but it usually yields acceptable results when carefully applied, and is hard to beat for its simplicity.

The EMA model must be applied with care, however, even for polysilicon. The data shown in Figure PA2.3 were acquired from a polysilicon film deposited by LPCVD at ~605°C. The optical constants of the polysilicon film were approximated by an EMA model containing three constituents—crystalline silicon, amorphous silicon, and voids.

In this case, the EMA model cannot accurately describe the optical constants of the polysilicon layer, even with three constituents. In our experience,

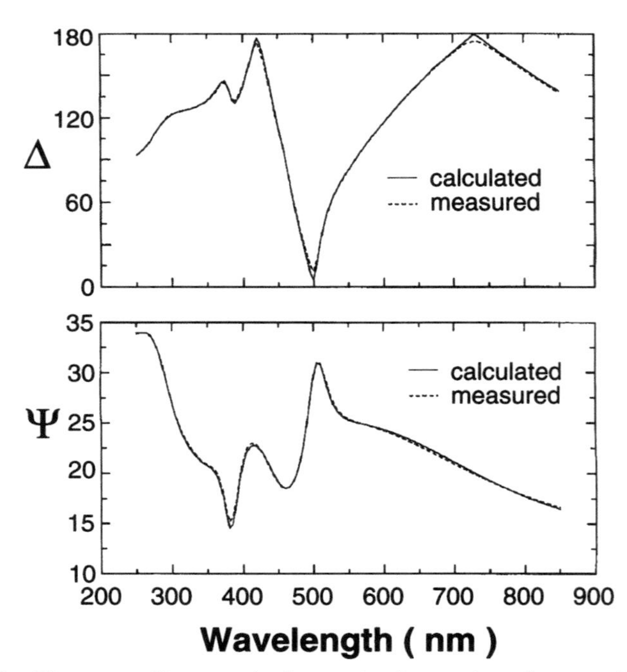

Figure PA2.2 The same ellipsometric data as in the previous figure, using a sophisticated dispersion model to describe the polysilicon layer optical constants.

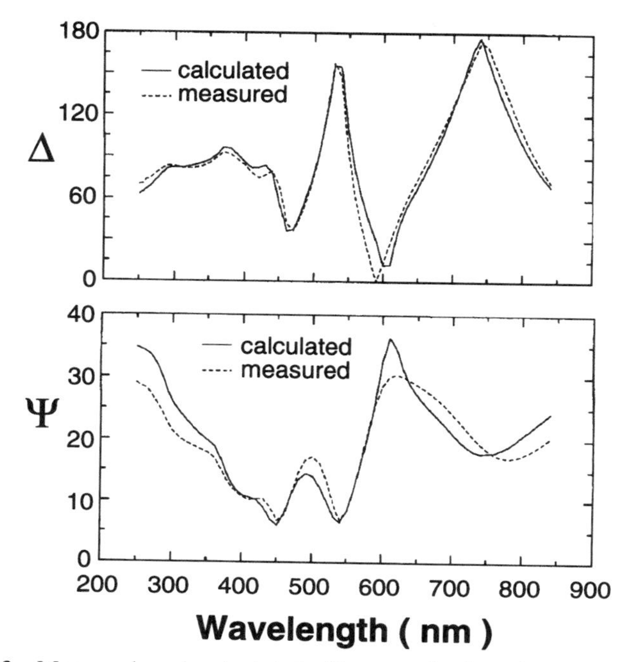

Figure PA2.3 Measured and calculated ellipsometric data for LPCVD polysilicon deposited at 605°C. Analysis was performed using a three-constituent EMA models for the polysilicon layer, mixing crystalline silicon, amorphous silicon, and voids.

the EMA model for polysilicon is most accurate when the polysilicon layer is either nearly crystalline or nearly amorphous, and tends to be less accurate in the intermediate region.

PA2.3 PECVD SILICON OXIDES

In this example we use the EMA model to describe a material which is not microstructurally a mixture of two known materials, but is stoichiometrically intermediate between two well-known compositions. For an off-stoichiometric silicon oxide, we can reasonably assume that the deposited film will have a stoichiometry somewhere in between silicon monoxide and silicon dioxide. Figure PA2.4 shows the optical constants of these two materials.

For this example we will analyze reflectance data from a 7400 Å film of BPSG deposited directly on silicon. The data were analyzed by varying the film thickness and the percentage of silicon monoxide in the film. The results are shown in Figure PA2.5.

Note that the experimental reflectance is matched fairly closely, but the oscillations of the experimental and calculated data become slightly out of

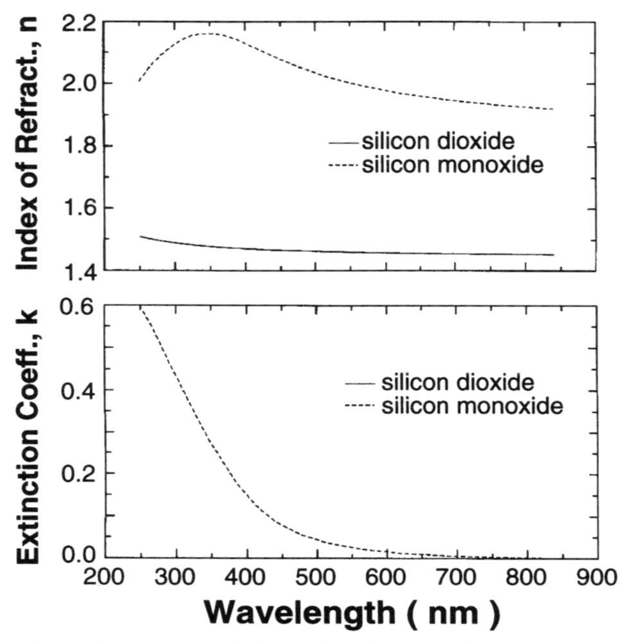

Figure PA2.4 Optical constants of silicon dioxide and silicon monoxide. Note that the extinction coefficient of silicon dioxide is zero over the spectral range shown.

phase towards the longer wavelength end of the spectrum. The EMA model places severe constraints on the dispersion of the film optical constants, and in this case is not quite able to get the dispersion of the BPSG index right. Mixing crystalline or amorphous silicon (rather than silicon monoxide) with silicon dioxide to model the BPSG film produces similar results. We will return to this data set further on to demonstrate the ability of more flexible dispersion models to model film optical constants more accurately. Nevertheless, the EMA type of model still provides useful information as to the stoichiometry of this film, and the effective silicon monoxide concentration from the EMA model may provide a sensitive monitor of the BPSG process in production.

PA2.4 SILICON OXYNITRIDES

We have reserved the most difficult of the EMA-type analyses for last. The silicon oxynitride material system is quite complicated, considering the number of choices available for mixing in EMA type models. We may choose to model an oxynitride film as a mixture of any of the following: silicon dioxide, silicon monoxide, silicon nitride, crystalline silicon, and amorphous silicon. In most cases, knowledge of the target composition for the film can

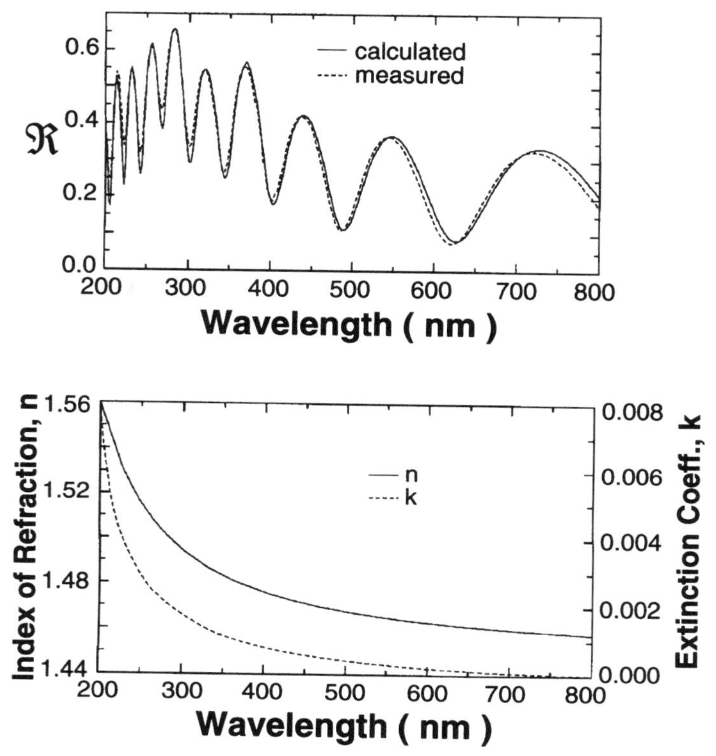

Figure PA2.5 Measured and calculated reflectance data from a thick BPSG film. Analysis yields a film thickness of 7560 Å and a 0.3% fraction of silicon monoxide in the film. The resulting optical constant for the BPSG film are also shown.

help in the choice of constituent materials, but in many cases we must resort to trying different combinations and using the mixture which best fits the experimental data.

The first example shown is a thick oxynitride film with composition somewhere between silicon dioxide and stoichiometric silicon nitride. After some experimentation, we found that the reflectance data from this sample was best fit by a three-constituent EMA mixture of silicon nitride, silicon dioxide, and crystalline silicon. The results are shown in Figure PA2.6.

This analysis yields a film thickness of 2959.7 Å, and fractions of 65% silicon nitride, 29.2% silicon dioxide, and 5.8% crystalline silicon in the film. Note that while the inclusion of crystalline silicon in the mixture helps to fit the data, it is highly unlikely that the fine structure observed in the resulting optical constants (caused by the sharp structure of the crystalline silicon optical constants) is real. We will also return to this data set further on in the prototypical analyses.

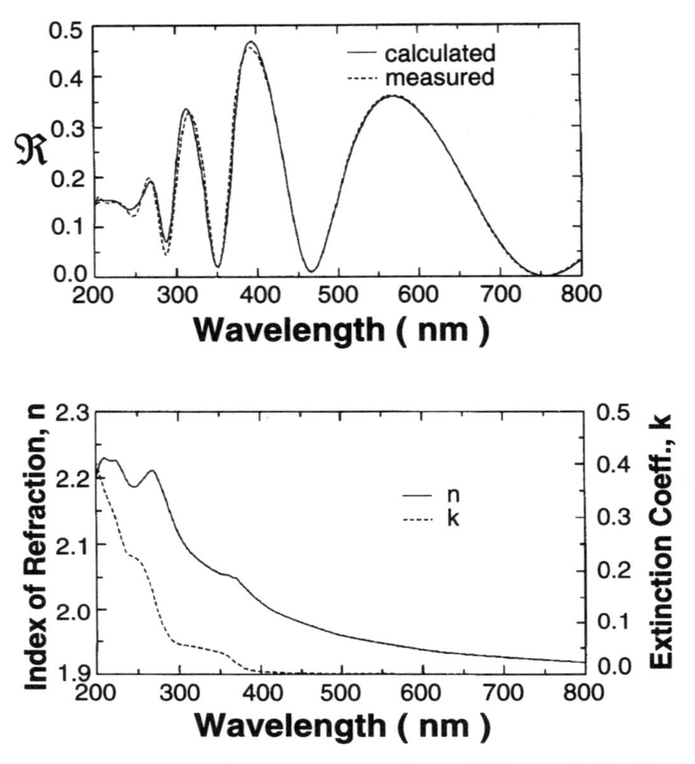

Figure PA2.6 EMA analysis of reflectance data from thick oxynitride film. The reflectance spectra are shown in the upper graph, and the resulting optical constants are shown in the lower graph.

The second example in this section will be a thin Si-rich oxynitride coating, commonly used as a bottom antireflection coating under photoresist layers. In this case, a mixture of silicon nitride, silicon monoxide, and crystalline silicon was found to provide the best fit to the ellipsometric data. The results are shown in Figure PA2.7.

In summary, EMA models can be very useful owing to their simplicity and natural physical interpretation. They are applicable to a wide range of different materials, but they must always be used with some care. The use of EMA models will generally cause the inclusion of some systematic error in the measurement results (as is the case with *all* dispersion models), but careful construction of the mixture to be used can help to minimize these errors.

A final word of caution. The EMA model is, as the name implies, a model. The fact that a given model gives a good fit implies that the optical constants are probably correct, but does not necessarily imply that the composition is correct. Recent (not yet published) studies by one of the authors (HGT) has shown that modeling PECVD nitrides as a combination of oxide and nitride

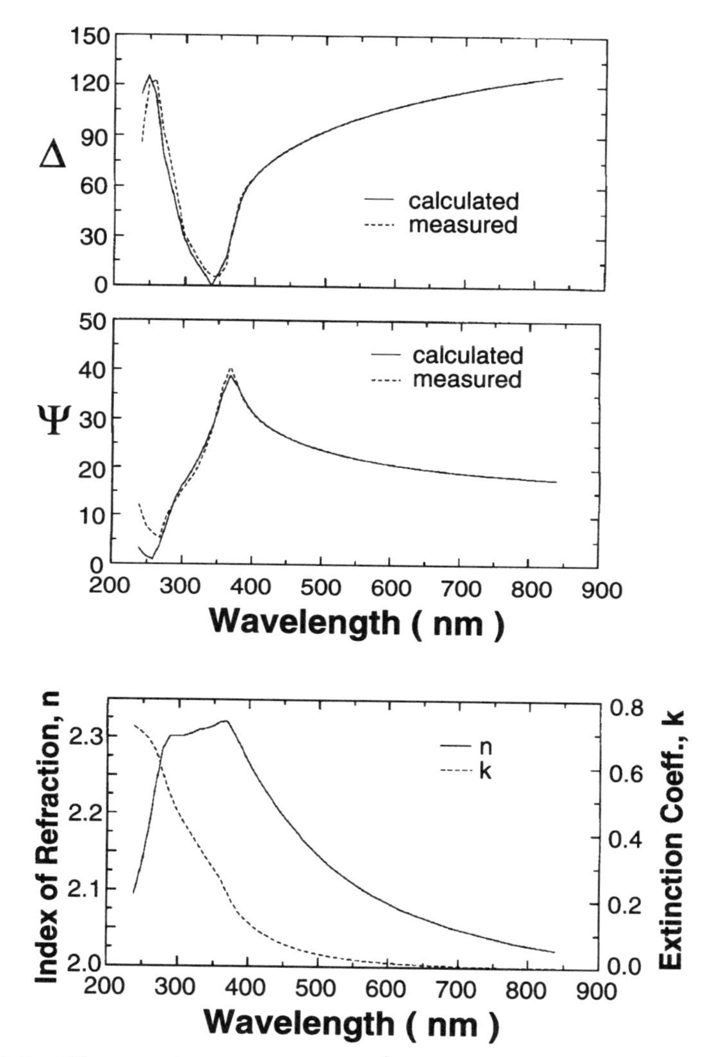

Figure PA2.7 Ellipsometric data for a 381.5 Å Si-rich oxynitride film on silicon. A mixture of silicon nitride with 78.3% silicon monoxide and 6.4% crystalline silicon yielded the fit shown.

always overestimates the amount of oxide. The reason for this is that in this model, the hydrogen which is present in PECVD oxynitrides is not taken into account. Hydrogen, when mixed with silicon nitride, lowers the index of refraction, which is the same result as mixing silicon dioxide with silicon nitride. The strongest statement which can be made with a good-fit EMA analysis is that the compositional results *are consistent with* the postulated mixture.

PA2.5 REFERENCES

1. D. E. Aspnes, J. B. Theeten, and F. Hottier, *Phys. Rev. B*, **20**, 3292 (1979).
2. H. G. Tompkins, *A User's Guide to Ellipsometry*, Academic Press, New York, 1993, Appendix B.

PROTOTYPICAL ANALYSIS No. 3

PECVD Silicon Nitride, Silicon Dioxide, and Photoresist on Silicon

PA3.1 SALIENT FEATURE

The salient feature of this prototypical analysis is that we will *use dispersion equations to describe the optical constants* of the film under study. In particular, we will examine the ubiquitous Cauchy model for the dispersion of the index of refraction of dielectric materials.

PA3.2 CAUCHY MODELING

The Cauchy dispersion model is a simple yet extremely powerful dispersion model for describing the index of refraction of dielectric and semiconducting materials. It tends to work best when the material under study shows little or no optical absorption in the spectral region of interest. The Cauchy dispersion model describes the dependence of the index of refraction on wavelength, as follows:

$$n(\lambda) = A + B/\lambda^2 + C/\lambda^4$$

where $n(\lambda)$ is the index of refraction, λ is the wavelength, and A, B, and C are parameters. Generally, the quantities to be determined are the thickness of the film and the values of A, B, and C.

Note that while A is dimensionless, B has dimensions of wavelength squared and C has dimensions of wavelength to the fourth power. This has caused immeasurable confusion when transporting Cauchy coefficients between different measuring tools, as several different sets of units are used by different manufacturers. For example, coefficient values calculated using wavelength units of angstroms will predict horribly different optical constants on a system which uses wavelength units of nanometers or micrometers.

Our first example will be a fairly thick oxide film. For comparison purposes, we will analyze the same set of data shown in Figure PA2.5 of the previous prototypical analysis. First, Figure PA3.1 shows what happens when we try to use the simple three-term Cauchy model shown above to analyze the data over the full spectral range 200–800 nm.

In this case, we have two choices. The simplest choice is simply to not fit the data in the UV region, and confine our analysis to the visible spectral range where the oxide film will be more transparent. Figure PA3.2 shows the results of this analysis.

Note the nearly perfect fit to the measured data, and that the index of refraction of the film is slightly higher than that of thermal oxide, as expected for a BPSG film. In some cases we may also wish to obtain the optical constants of the BPSG film in the UV. In this case, we must also model the absorption of the oxide film in the UV in order to obtain the most accurate results. There are a number of useful models for absorption *tails* used in conjunction with Cauchy index models, one of the more popular of which is a simple exponential function (discussed in Section 3.5):

$$k(\lambda) = C_1 e^{C_2(E - E_b)}$$

where $k(\lambda)$ is the extinction coefficient, and C_1, C_2, and E_b are fit parameters. Using this function and fitting for these parameters in addition to the three Cauchy parameters for the index of refraction yields the results shown in figure PA3.3.

The key to successful use of Cauchy dispersion models is to apply them only over spectral ranges where the film they are used to describe is transparent, or at least has a small absorption tail for which the exponential function (or whatever function is being used for the extinction coefficient) is capable of

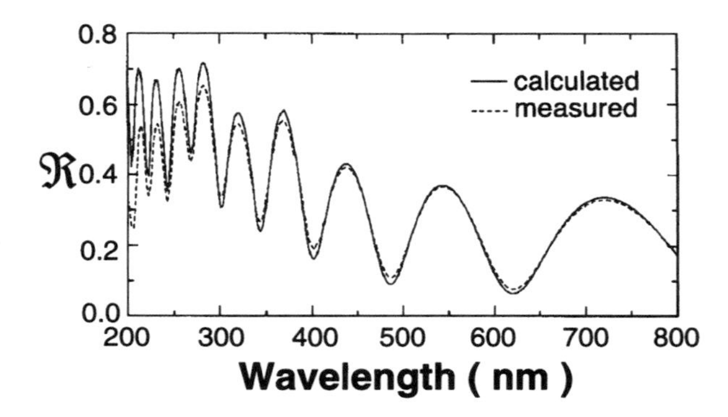

Figure PA3.1 Analysis of reflectance data from thick oxide using a Cauchy model for the oxide index of refraction. Note the poor fit in the UV due to absorption in the oxide film.

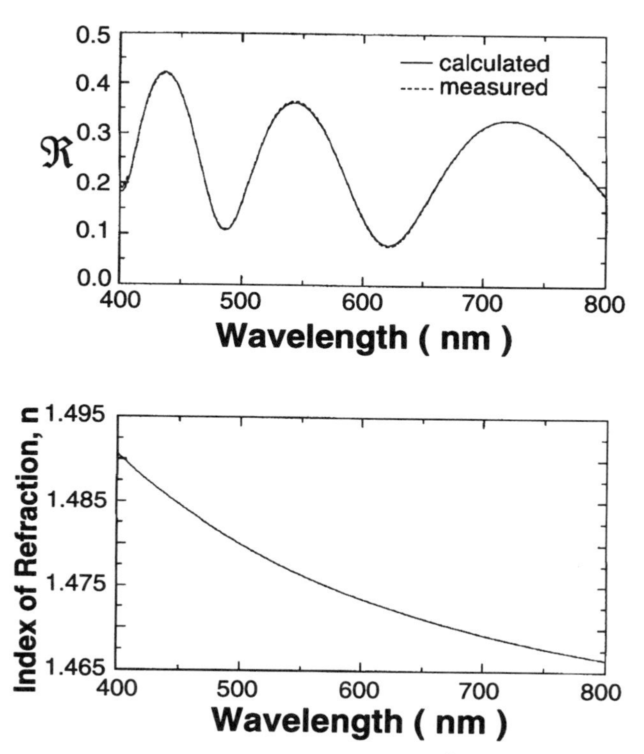

Figure PA3.2 Results of Cauchy analysis for 7442.4-Å thick oxide film using a three-term Cauchy model and fitting data only in the visible. The resulting Cauchy parameters are $A = 1.457$, $B = 0.0066$, and $C = -0.00018$, with the wavelength units in micrometers.

accurately modeling the absorption of the film. To illustrate this, our next analysis will be of a photoresist film on silicon.

Figure PA3.4 shows the results of a standard Cauchy analysis of a thick photoresist film on silicon. A two-term Cauchy model with no absorption was used in order to determine the film thickness and index of refraction.

Note that in this analysis, the measured data is only fit over the spectral range 600–800 nm, and that the match between the measured and calculated data is excellent. The film thickness obtained from the above analysis is 14066 Å.

Figure PA3.5 shows what happens if we try to perform the same analysis over a slightly broader spectral range. Note that this is a dyed resist, and we have included the exponential absorption tail in an attempt to get a better fit to the data.

This analysis yields a film thickness of 13802 Å, which is 264 Å less than that obtained when fitting from 600 to 800 nm only. The dispersion of the resist optical constants from 550 to 600 nm is sufficiently complicated that the

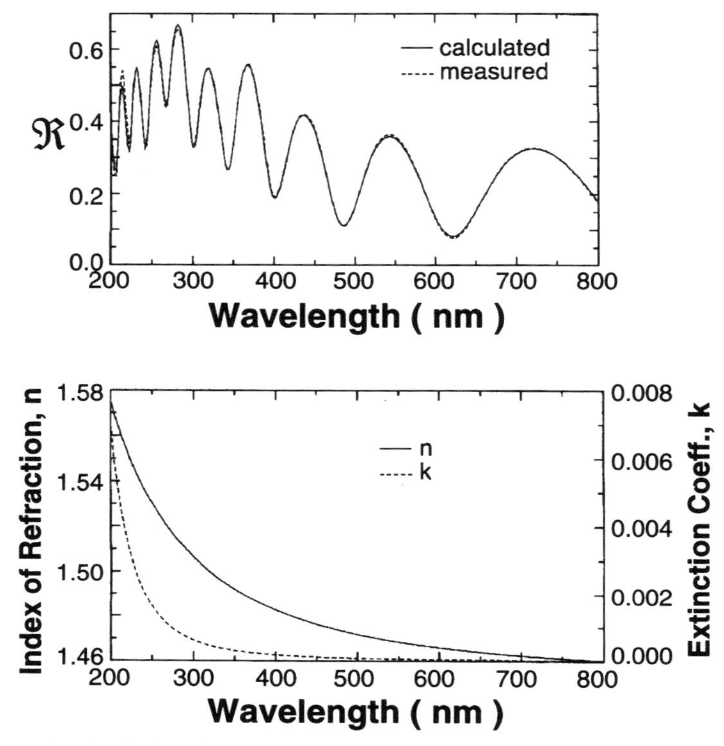

Figure PA3.3 Analysis of reflectance data using a three-term Cauchy model with an exponential absorption tail. Note the much better fit to the experimental data in the UV region of the spectrum.

Cauchy model with an exponential absorption tail is simply not capable of reproducing them. Any dispersion model must be used with care, and only in cases where the given model has a reasonable chance of actually reproducing the real optical constants of the film.

The final example in this section illustrates one of the most powerful and commonly used applications of the Cauchy model for ellipsometric data analysis. In this example, we wish to determine the optical constants of an organic antireflection coating (ARC) which has been deposited on a silicon wafer. Ellipsometric data were acquired from this sample at three angles of incidence from 200 to 800 nm. Note in the following figures that an instrument with a compensator was used for the data collection; hence delta ranges from zero to 360° rather than from zero to 180°

This analysis proceeds in two steps. First, we use the simple Cauchy model to determine the thickness of the ARC film. This consists of finding a portion of the spectral range in which the film is transparent so that the Cauchy model can accurately describe the ARC film index of refraction. Once the film thick-

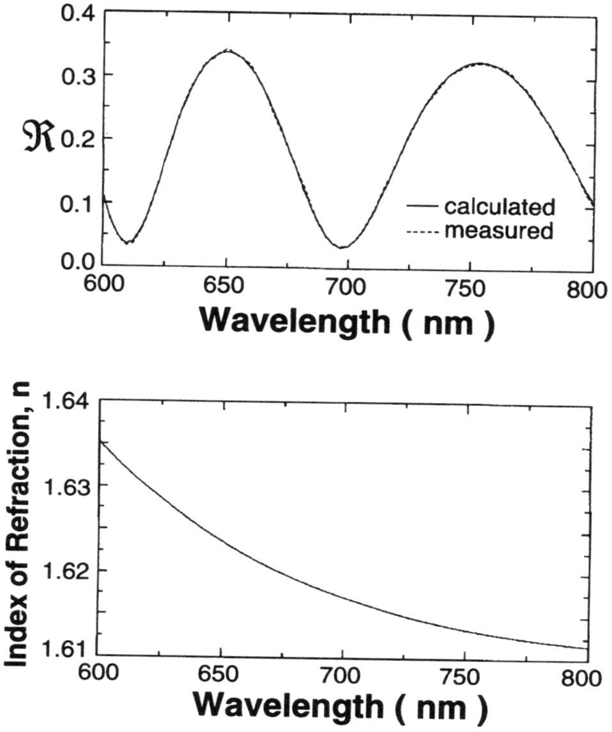

Figure PA3.4 Analysis of thick photoresist on silicon using a two-term Cauchy model with no absorption for the resist optical constants. Top, measured and calculated reflectance data; bottom, the resulting index of refraction for the photoresist.

ness has been precisely determined by this analysis, we can take the thickness as a fixed parameter and directly invert the ellipsometric data in order to obtain the optical constants at each measurement wavelength. This is a very powerful analysis technique for films which exhibit complicated optical constant dispersion.

We initially fit the ellipsometric data using a nonabsorbing Cauchy model with three terms, fitting for the film thickness and the three Cauchy parameters. Additionally, we found that the inclusion of a surface roughness layer (modeled using an EMA of 50% of the film material and 50% voids) significantly improves the fit to the data. A good approach to determining the appropriate spectral range for this analysis is to start with a large spectral range (say 300–800 nm), perform the Cauchy fit, then shorten the spectral range (to 350–800 nm, for example) and repeat the fitting procedure. When shortening the spectral range fails to improve the final MSE and the fit to the experimental data is good, it is probable that the film is transparent in this spectral range. Table PA3.1 shows what happened when we applied this procedure to the data from the ARC coating.

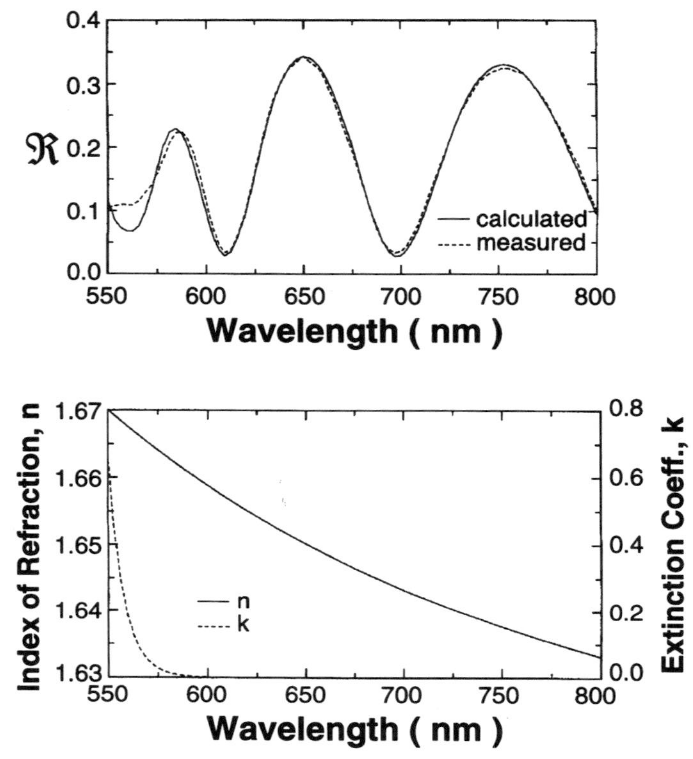

Figure PA3.5 Extending the wavelength range for the analysis of the sample shown in the previous figure.

We conclude that over the spectral range 500–800 nm, we have basically achieved a perfect fit to the experimental data, and the film is transparent in this range. The results of this fit are shown in Figure PA3.6.

This analysis yields a film thickness of 2012.4 Å, with 30.5 Å of surface roughness. We then use these thicknesses and fit the ellipsometric data at

TABLE PA3.1 MSE for Various Spectral Ranges

Spectral Range (nm)	MSE
300–800	116.7
350–800	103.5
400–800	16.6
450–800	2.1
500–800	0.65
550–800	0.57
600–800	0.53

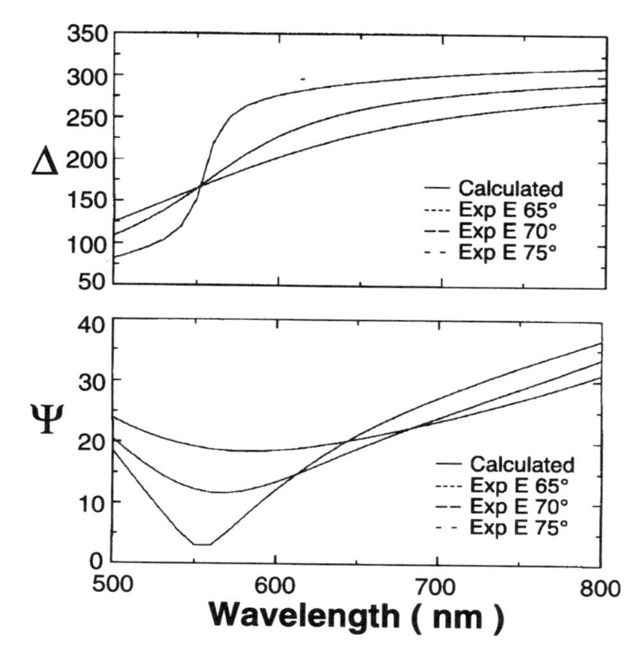

Figure PA3.6 Measured and calculated ellipsometric data from an organic ARC coating on silicon, modeled with a three-term Cauchy model with no absorption.

each measurement wavelength to determine the optical constant values for the film at that wavelength. The results of this analysis are shown in Figure PA3.7.

It is also interesting to note that the film is actually transparent for wavelengths greater than about 420 nm, even though our Cauchy fitting procedure did not stabilize until about 500 nm. This indicates that the Cauchy model could not accurately reproduce the dispersion of the index of refraction of the ARC coating from 420 to 500 nm. Examination of Figure PA3.7 shows that the index of the ARC coating begins to increase sharply as the wavelength decreases in this spectral range, and the simple three term Cauchy model could not accurately model this dispersion. We conclude that not only should the film under study show no absorption in the spectral range to be fit for the Cauchy model to work well, but also that the index of the film should not show strong dispersion in this range.

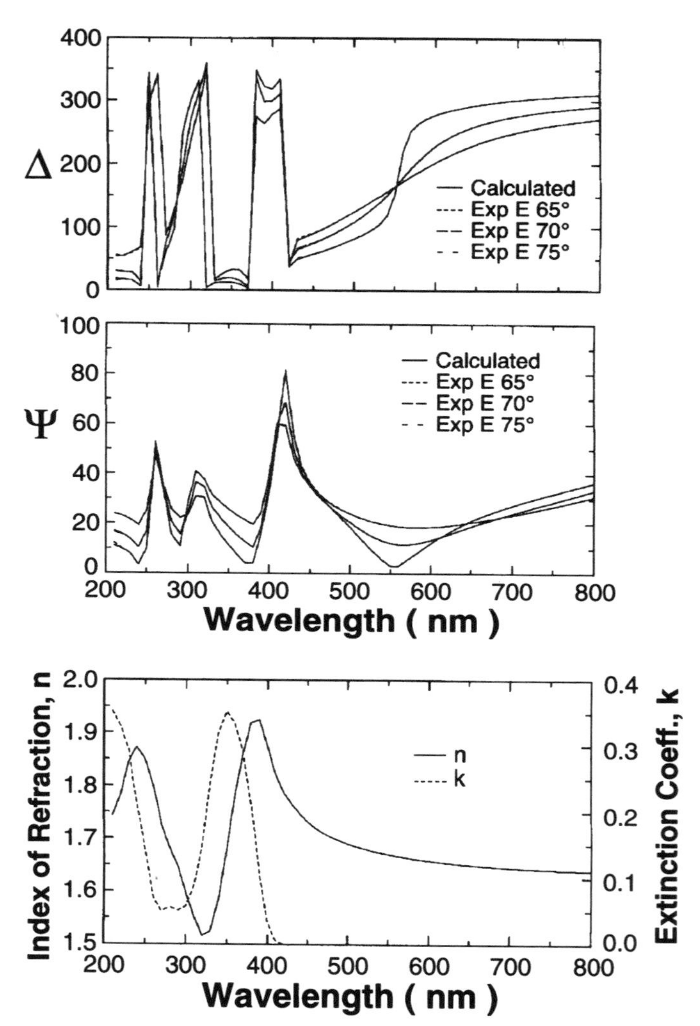

Figure PA3.7 Results of direct calculation of optical constants of the ARC film from the measured ellipsometric data using the film and roughness obtained from the Cauchy analysis. Note the complicated dispersion of the optical constants of the film below 400 nm.

PROTOTYPICAL ANALYSIS No. 4

LPCVD Polysilicon and Amorphous Silicon

PA4.1 SALIENT FEATURE

The salient feature of this prototypical analysis will be *the use of simple dispersion models to describe the optical constants of process-variable films in multilayer structures*. We will also expand on the use of some more sophisticated dispersion models to describe polysilicon.

PA4.2 THE PROBLEM

This seemingly simple structure has caused enormous problems in semiconductor metrology for many years. This is owing to the complicated nature of the polysilicon optical constant spectra, and the tendency of the polysilicon optical constants to change significantly with small changes in the deposition or anneal process. Metrology tools which relied on fixed optical constants for the polysilicon layer in order to determine the film thickness tended to suffer from inaccuracies due to small changes in the polysilicon optical constants, such as those caused by nonuniform temperature profiles in a polysilicon reactor. Also, the inevitable presence of roughness at the surface of polysilicon films further complicates the analysis of the data, particularly in the UV portion of the spectrum. This roughness tends to scale with the thickness of the polysilicon film, so that it may be barely noticeable for thin films, but can be severely problematic for thicker polysilicon films. The roughness of amorphous silicon is significantly smaller than for silicon deposited in the polycrystalline form.

PA4.3 THE OPTICAL CONSTANTS

Before starting to analyze polysilicon, it is helpful to understand the behavior of its optical constants. (This is true for any material, but particularly true for polysilicon.) Figure PA4.1 shows the optical constants of LPCVD polysilicon films as a function of deposition temperature.

It is obvious from this figure that the optical constants of polysilicon are extremely strongly dependent on the deposition temperature. The middle temperature range of about 590–620°C is particularly problematic, as the polysilicon optical constants drop below those of crystalline silicon. The EMA model has a very difficult time modeling the polysilicon optical constants in this temperature range, as seen in prototypical analysis No. 2.

Completely different behavior is observed if an amorphous silicon film is deposited and then annealed for differing periods of time, as shown in Figure PA4.2.

In this case, the polysilicon optical constants vary systematically with anneal from a-Si to crystalline silicon. It is much more reasonable to approximate this material as a mixture of amorphous and crystalline silicon, and indeed the EMA model works reasonably well in this case, particularly when the film is nearly amorphous or highly crystalline.

PA4.4 USING THE EMA MODEL

For our first example we will analyze a thin polysilicon film on a 1000 Å thick thermal oxide on silicon. This is a very common structure, as for many years the standard procedure for polysilicon analysis was to deposit the polysilicon film on a 1000 Å thermal oxide monitor wafer. Our initial analysis will be made using an EMA model to mix crystalline and amorphous silicon to describe the polysilicon layer. We will fit for the amorphous silicon fraction in the polysilicon layer and the thicknesses of the thermal oxide, polysilicon, and roughness layers. The results are shown in Figure PA4.3.

This analysis yields thicknesses of 922.4 Å, 1157.4 Å, and 21.2 Å for the thermal oxide, polysilicon, and surface roughness layers, respectively, and an amorphous silicon fraction of 25.0% in the polysilicon layer. Note that the match between the measured and calculated data is not particularly good.

PA4.5 USING THE LORENTZ OSCILLATOR DISPERSION MODEL

In order to obtain a better match between the measured and calculated data, it is necessary to employ a somewhat more sophisticated dispersion model for the polysilicon optical constants. For this example, we will use a Lorentz oscillator dispersion model (described in a previous chapter) to describe the polysilicon optical constants.

It is generally a good idea to convince oneself that the dispersion model to be used is capable of accurately reproducing the optical constants of the material under study. Also, since we do not already know the optical constants of the polysilicon layer, we need to obtain some sort of initial guesses for the

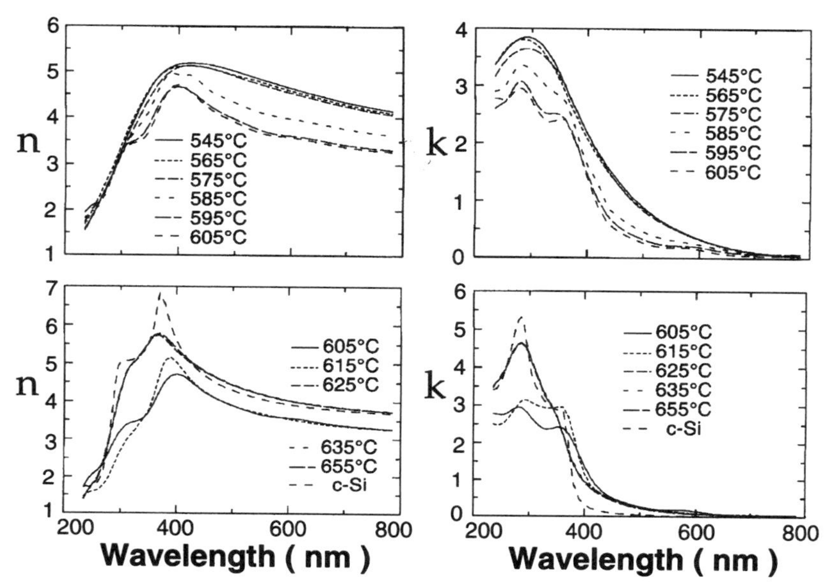

Figure PA4.1 Polysilicon optical constants as a function of LPCVD deposition temperature. The top two figures show the n and k spectra for temperatures from 545°C to 605°C, while the bottom two show the spectra for temperatures from 605°C to 655°C.

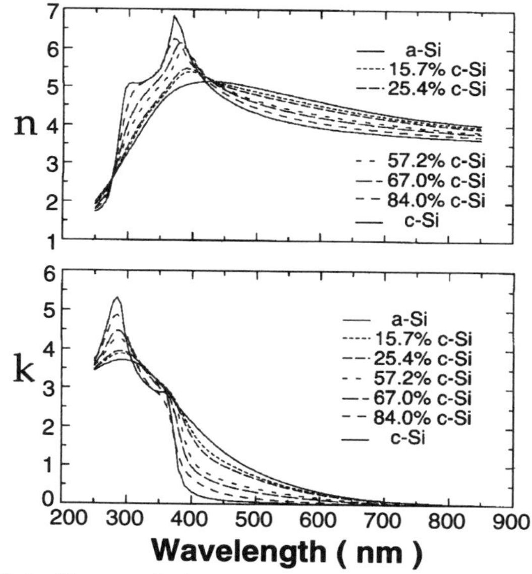

Figure PA4.2 Polysilicon optical constants as a function of percentage crystallinity, determined by annealing time and temperature. All films were deposited as amorphous silicon and then annealed. The percent crystallinity values were obtained from an EMA analysis.

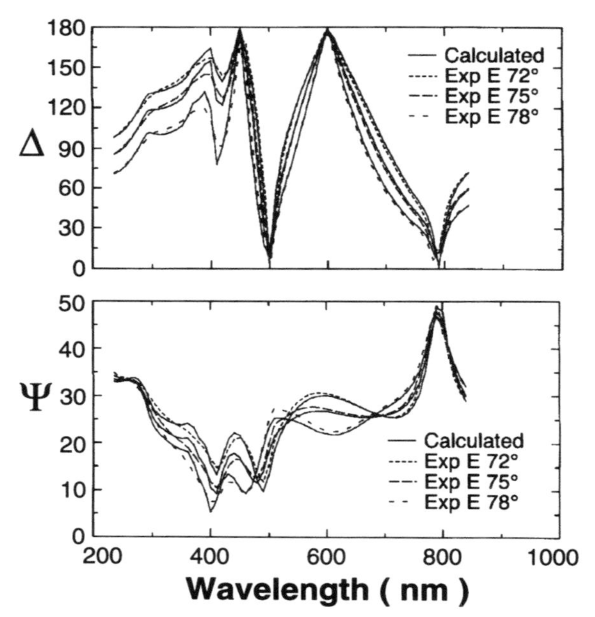

Figure PA4.3 Analysis of polysilicon film on thermal oxide using a two-constituent EMA model for the polysilicon layer.

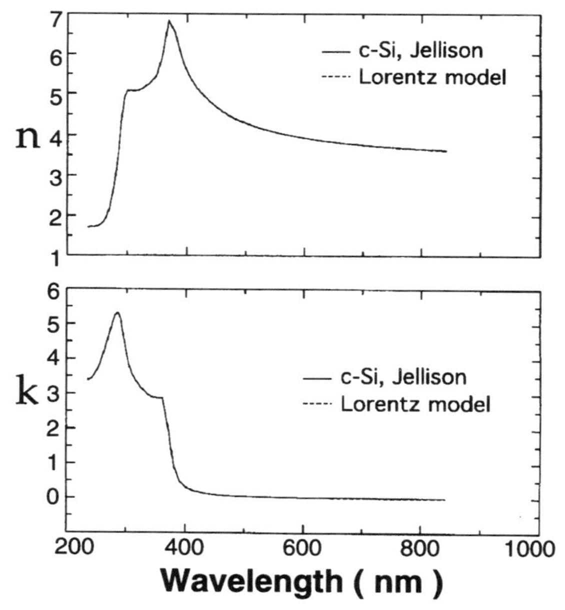

Figure PA4.4 Lorentz oscillator fit to c-Si constants which were developed by Jellison.[1]

number of oscillators and the initial values of the oscillator parameters. To do this, we fit the Lorentz oscillator model directly to the optical constants of crystalline silicon.[1] We found that a reasonable fit could be obtained with a seven-oscillator model provided we allowed the amplitude of two of the oscillators to be negative. The fits to the crystalline silicon optical constants are shown in Figure PA4.4.

We then use this Lorentz oscillator parameterization as the starting point to analyze the polysilicon data. We initially fit for the thermal oxide, polysilicon,

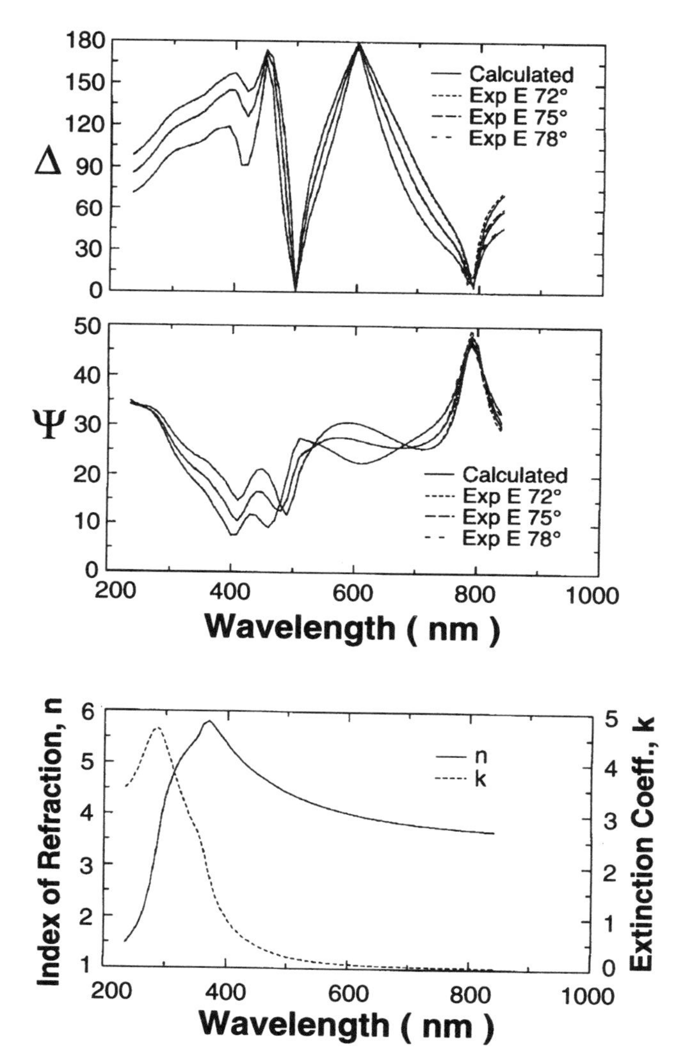

Figure PA4.5 Measured and calculated ellipsometric data for polysilicon on oxide on silicon. The resulting optical constants of the polysilicon film are also shown.

and roughness layer thicknesses and the amplitudes of all seven oscillators. We then added the broadenings and center energies of the seven oscillators to the set of fit parameters in order to fine-tune the analysis. We found that when fitting for all of the oscillator parameters there was a very strong correlation between the thermal oxide layer thickness and the Lorentz oscillator parameters, and as a result we were forced to fix the thickness of the thermal oxide layer at its nominal value, 1000 Å. The results of this analysis are shown in Figure PA4.5.

This analysis yields thicknesses of 1168.8 Å for the polysilicon layer and 19.8 Å for the surface roughness. Note that the match between the measured and calculated data is much better, although still not perfect (particularly at the longer wavelengths). The use of the Lorentz oscillator layer provides a much better fit to the data at the expense of a much more complex parameter set for the dispersion model for the polysilicon.

PA4.6 REFERENCES

1. G. E. Jellison, Jr., *Optical Materials*, **1**, 41 (1992).

PROTOTYPICAL ANALYSIS No. 5

Substrate Optical Constant Determination

PA5.1 SALIENT FEATURE

The salient feature of this prototypical analysis is *the accurate determination of substrate optical constants*. This can be critically important for the further accurate characterization of films deposited on these substrates. The best technique for this type of analysis is variable angle of incidence spectroscopic ellipsometry, as it provides the ability to calculate the optical constants of the substrate directly at each measured wavelength. Reasonable measurements of substrate optical constants can still be performed by other techniques through the use of dispersion modeling.

PA5.2 ALUMINUM

While bulk aluminum substrates are not commonly encountered, thick films of aluminum are common in the semiconductor industry. Aluminum films more than ~500 Å thick are opaque and therefore function effectively as the substrate for optical characterization. A very common problem, for example, is the accurate characterization of a thick aluminum film such that oxide films deposited on the aluminum can be measured accurately.

The simplest means of analysis of ellipsometric data from a substrate (or very thick metal film, in this case) is to calculate the optical constants of the substrate directly from the measured ellipsometric data at each measurement wavelength. There are two optical constants (n and k), and two ellipsometric parameters are measured at each wavelength (psi and delta); therefore it is possible to calculate n and k from a single ellipsometric measurement. Any error in the measured psi and delta, however, will map directly into the calculated n and k values, so it is generally better to acquire ellipsometric data at more than one angle of incidence so that the optical constants are overdetermined and the measurement noise will be averaged out.

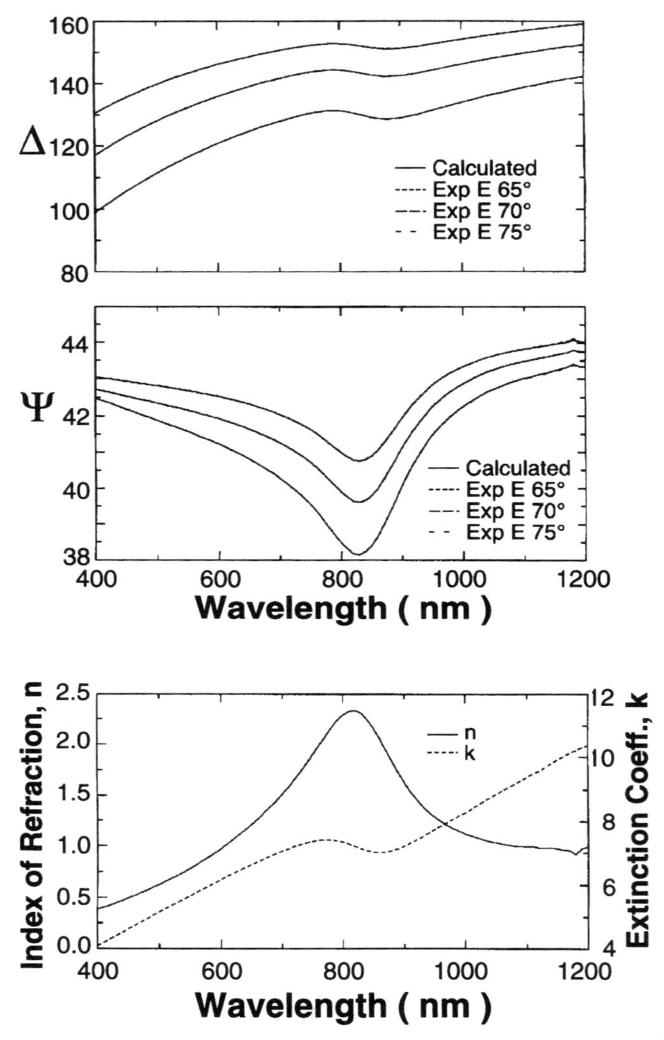

Figure PA5.1 measured and calculated ellipsometric data at three angles of incidence for a thick aluminium film. The optical constant spectra obtained for the film from this analysis are also shown.

Figure PA5.1 shows the results of the analysis of ellipsometric data acquired at three angles of incidence from a thick aluminum film on a silicon substrate. The data were fit by varying the values of n and k at each measurement wavelength.

Note that this analysis assumes that the surface of the aluminum film is perfectly smooth and is not oxidized. It is probable that neither of these assumptions is true, and the accuracy of the optical constants we obtained will suffer from the presence of roughness and/or oxidation of the sample. We cannot determine the extent of roughness or oxidation (as well as the aluminum optical constants) from the ellipsometric data, as we have already obtained a perfect fit to the experimental data by varying only the aluminum optical constants. We can, however, include a roughness or oxide layer of fixed thickness and obtain the aluminum optical constants. This is common practice when a reasonable estimate of the oxide or roughness thickness can be made. However, for production applications we most often use the "effective" optical constants of the substrate obtained without surface roughness or oxides, particularly if the substrate "sees" atmosphere before the film deposition.

PA5.3 GLASS

The final example in this section will be the characterization of a bare soda-lime glass substrate. This analysis can be somewhat difficult owing to the reflection

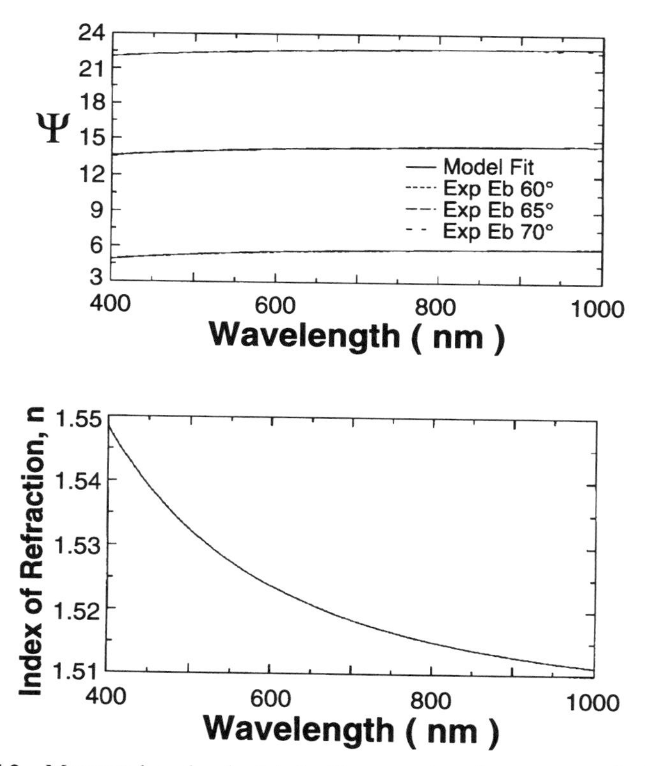

Figure PA5.2 Measured and calculated ellipsometric psi spectra, and glass index of refraction obtained from this fit.

of the light beam from the back surface of the substrate. In this case, we have obtained ellipsometric data at three angles of incidence from the substrate, as well as intensity transmission data at normal incidence through the substrate. The analysis of this data proceeds in two steps. First, we fit the ellipsometric psi data only in order to determine the index of refraction of the glass, using a Cauchy parametrization for the glass index of refraction. We do not use the delta data as it is very strongly affected by the back-surface reflection. For a perfect dielectric, that is, $k = 0$, the value of delta would be either zero or 180°. Rotating-element ellipsometers which do not have a compensator can be very inaccurate at these values of delta. The computational result of measurement deviations from zero or 180° is a nonzero value of k, which we know to be untrue. The results of this analysis are shown in figure PA5.2.

This analysis yielded Cauchy parameters for the glass of $A = 1.504$ and $B = 0.00708$, with C fixed at zero. The second step of the analysis procedure is to fit the transmission data in order to determine the glass extinction coefficient in the UV. The results of this analysis are shown in Figure PA5.3.

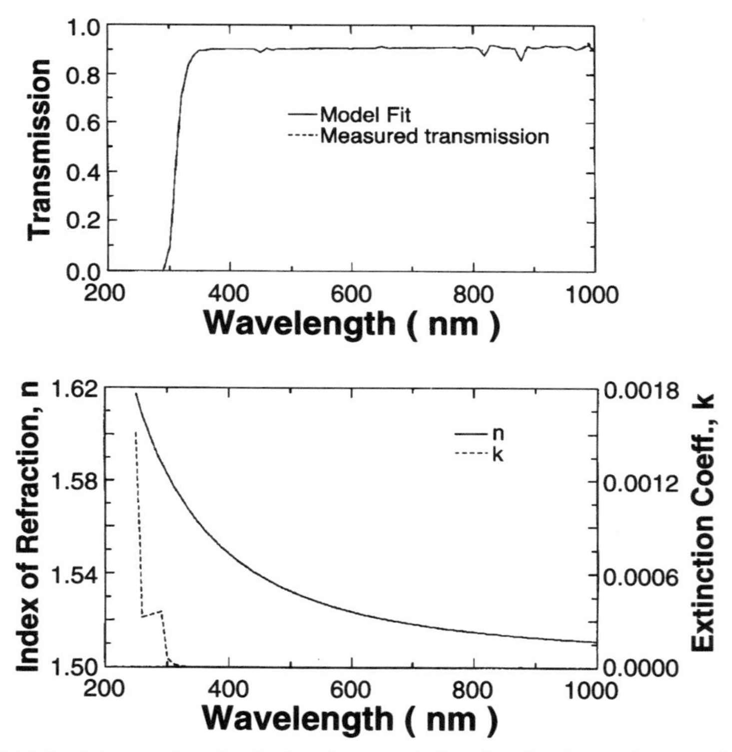

Figure PA5.3 Measured and calculated transmission for the glass substrate. The optical constants of the glass substrate are also shown.

Note that the extinction coefficient values obtained for wavelengths where the transmission is zero are somewhat meaningless, as any value of the extinction coefficient greater than the value required to extinguish the light beam will yield an equally good fit. This procedure yields reasonably accurate optical constants for glass substrates, without requiring destructive modification of the sample.

PROTOTYPICAL ANALYSIS No. 6

Analysis of Films on Transparent Substrates

PA6.1 SALIENT FEATURE

The salient feature of this prototypical analysis is that *the substrate is transparent*. Examples of this type analysis are films on glass substrates. A transparent substrate gives some advantages and some disadvantages.

PA6.2 THE STRUCTURE OF THE EXAMPLE

In order to illustrate the unique features of this type of analysis, we use an example of amorphous silicon on a soda-lime glass substrate. The film was slightly less than 5000 Å thick and the optical constants of the film are shown in Figure PA6.1. The soda-lime glass has an index of refraction spectrum which is similar to that of thermally grown SiO_2, except that it is slightly higher. In addition, the extinction coefficient for soda-lime glass below a wavelength of 3500 Å is nonzero, although it is small enough to be ignored for most normal ellipsometry and reflectometry. Transmission, reflection, and ellipsometry measurements were made on this sample. In addition, the back surface was abraded at a nearby location. Reflectance and ellipsometry measurements were made on this location. In this case, all of the light which reached the rear surface was scattered and did not return to the measurement instrument.

PA6.3 TRANSMISSION

Many ellipsometers are also equipped to make transmission measurements. For most semiconductor applications, this is of no value. The extinction coefficients for Si and GaAs are small but nonzero, and for a typical wafer visible light does not reach the rear surface. For other technologies (e.g., flat panel

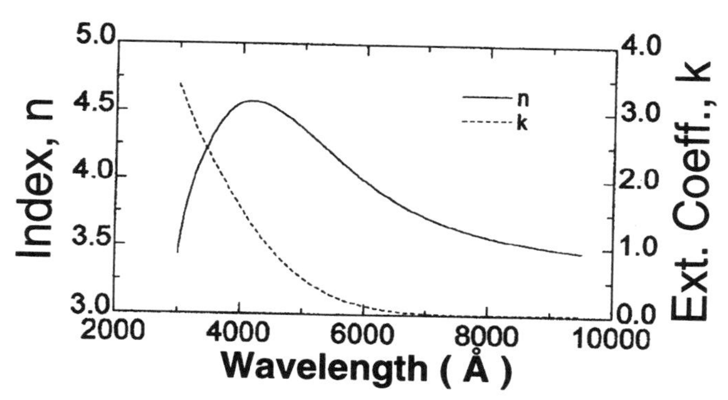

Figure PA6.1 Optical constants for the example film.

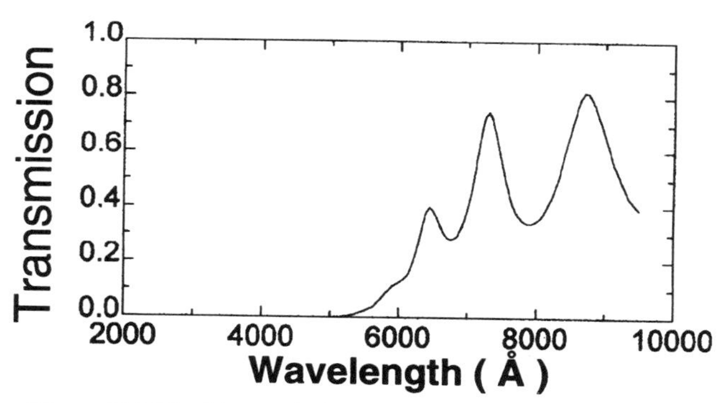

Figure PA6.2 Transmission spectrum for the example film.

display), when the substrate is transparent, a transmission measurement provides useful information. Figure PA6.2 shows the transmission spectrum for our example film. From this figure, we can see that our film is opaque for wavelengths below 5000 Å, and that fact, along with the optical constant spectrum, will provide a rough estimate of the thickness. Constructive and destructive interference between the top and bottom of the amorphous Si give the maxima and minima shown in the figure. We defer until later a discussion of the reflection from the back of the substrate. For a single film for which optical constants are available, thickness can be determined from this transmission spectrum alone.

PA6.4 REFLECTION

For an opaque substrate, the light reflects from the various surfaces in the film structure, as indicated in Figure PA6.3 as beam 1, but the light which is

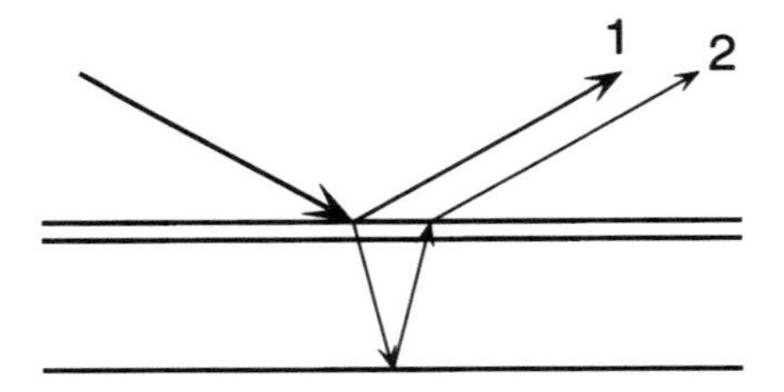

Figure PA6.3 Light reflecting from the film side of the substrate (1) and from the rear side of the substrate (2). The beams are shown separated, but in actual practice, both beams re-enter the instrument and contribute to the, measured light.

transmitted into the substrate is absorbed and does not reach the lower interface. For a transparent substrate, however, the light reflects from the back surface of the substrate and returns to the film structure. Some of the light is transmitted through the film and becomes part of the light which is measured. This is indicated in the figure as beam 2. Although in the figure the beams are shown separated, in actual practice both contribute to the measured beam. For the reflection from the rear surface, the distance traveled is such that all coherence is lost; hence this light does not interfere either constructively or destructively, but instead simply adds intensity. This can be handled mathematically[1] by modifying the equation for the total reflection coefficient (Eq. 2.18).

For our particular example, the amorphous silicon has an index of refraction which is rather high compared with that of either air or glass, and much of the light is reflected from the front and back surfaces of the silicon film. At normal incidence, the amount of light which reflects from the rear surface is only a few percent of the incident light and the effect on a reflectance spectrum is small. Figure PA6.4 shows a reflectance spectrum for the example when the

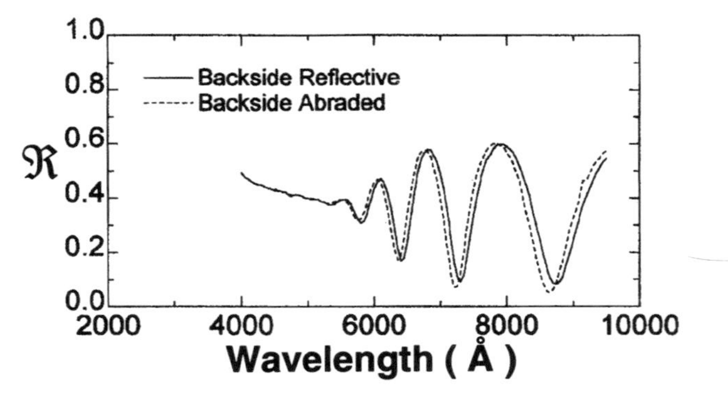

Figure PA6.4 Reflectance spectrum for the example film. Solid line, the sample where the reflection from the rear surface is included. Dashed line, the location which had the backside abraded and will have no contribution from the rear surface. The slight offset is due to a slight difference in film thickness (~ 50Å) and is not due to the rear surface.

rear surface is unaltered, and a spectrum where the rear surface is abraded. In many cases for commercial reflectance instruments, the reflection from the back surface is simply ignored. A high N/A objective reduces this effect and is commonly used in flat-panel display production.

PA6.5 ELLIPSOMETRY

For ellipsometry measurements at a high angle of incidence, the effect of the backside cannot be ignored. The p-wave contribution of the rear surface reflection will be significantly different from the s-wave contribution (as suggested in Figure 2.6). As shown in Figure PA6.3, the beam designated as 1 will carry the information about the film. The beam designated as 2 will add intensity to each component, but does not contain any useful information.

Various schemes are used to remove this useless light from the measurement process. The most common method is to abrade the rear surface, giving a surface which scatters the light rather than reflecting it back into the instrument. This method works very well in a laboratory environment where destructive methods are often used. It is not very practical in a microelectronics environment, however, where glass particles are frowned on and the sample

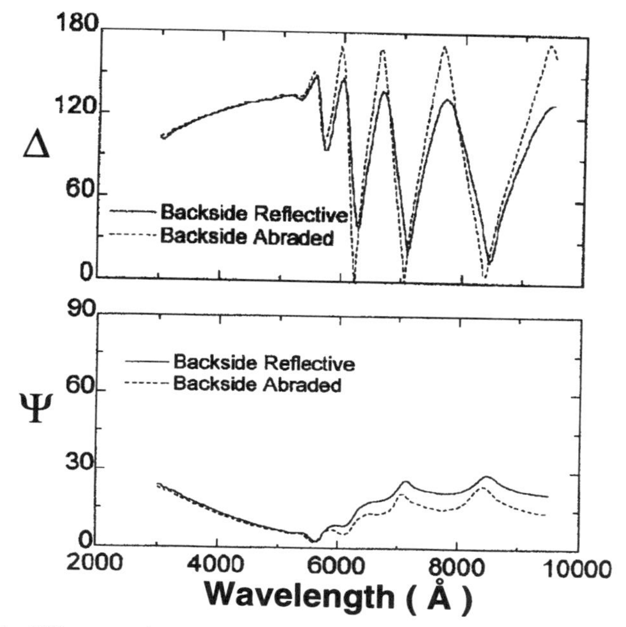

Figure PA6.5 Ellipsometric spectra for the example film. Dashed line, a sample where the backside has been abraded. This is normal ellipsometry. Solid line, a sample where the backside is reflective.

is often supposed to become part of a product. In some cases, the software for the measuring instrument can take the rear surface reflection into account if instructed to do so.

Figure PA6.5 shows ellipsometric spectra for a sample with a reflective rear surface and for a sample with an abraded rear surface. The abraded sample is the "normal" situation, whereas the backside reflection of the reflective sample must be taken into account. The effect of the light from the rear surface appears to be to decrease the "intensity" of the undulations in the delta curve and to increase the value of psi.

PA6.6 REVERSE ELLIPSOMETRY

Occasionally, it is advantageous to bring the light to the film from the rear, through the substrate. An example of this might be a stack of films where the middle one is opaque for most of the wavelength range.[2] The films between the opaque layer and the substrate will be "hidden" from the analysis. When the substrate is transparent, reverse ellipsometry can be used. The optical configuration is shown in Figure PA6.6.

The light strikes the substrate and some of it (beam 1) reflects. This light carries no useful information, but cannot be separated from the useful light. Some of the light enters the substrate, travels to the opposite side, and interacts with the film stack (shown as a single layer in the figure). Some of the light reflects from the film stack back into the substrate, travels to the opposite side and emerges (beam 2). This beam carries the information about the film stack. This is particularly useful for determining the optical constants of reactive metal films.

Figure PA6.7 shows the ellipsometric spectra for our example film when the reverse configuration is used. Also shown for comparison is the spectra from the front side for the abraded sample.

The effect of the first reflection from the substrate (the back) again causes the value of delta to be sharply reduced and the values of psi to be increased slightly. The decrease in both delta and psi for wavelengths below 3500 Å is due

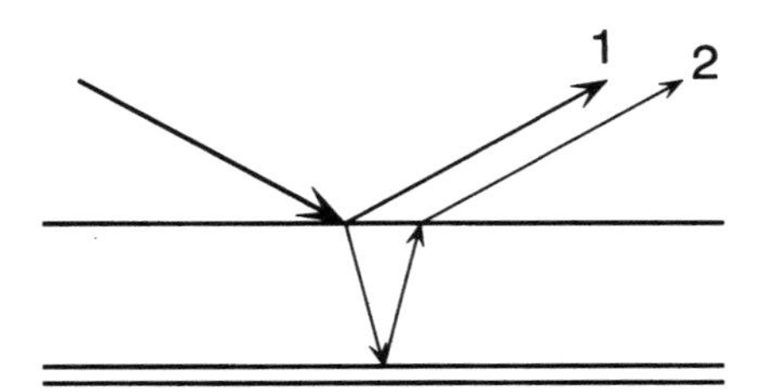

Figure PA6.6 The light beam configuration for reverse ellipsometry. In this case, the light travels through the substrate and interacts with the film stack. Beam 2 contains information about the film stack, whereas beam 1 contains no useful information.

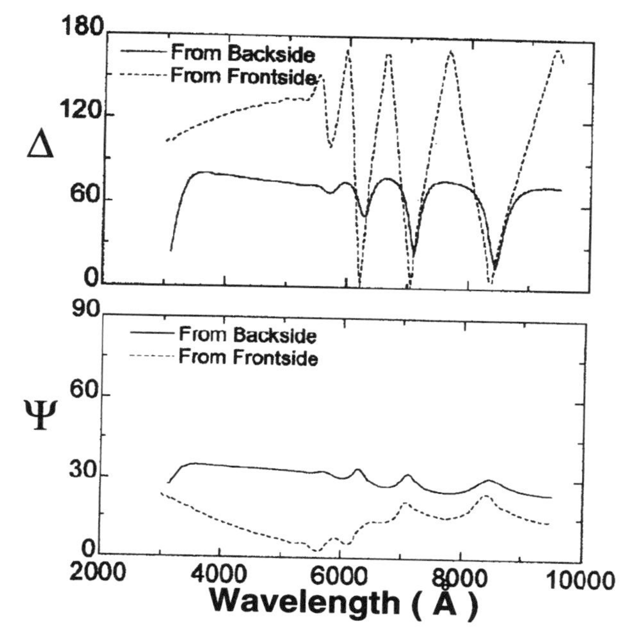

Figure PA6.7 Ellipsometric spectra for the example film when the reverse configuration is used. Also shown for comparison is the spectra for the abraded sample.

to a small nonzero value of the extinction coefficient of the soda-lime glass substrate. The value is small enough that it can be ignored for normal ellipsometry, but has an effect when the light passes through a millimeter of glass.

PA6.7 SUMMARY

Transparent substrates have some advantages and some disadvantages. The advantages are:

- transmission measurements can be made;
- in some situations, reverse ellipsometry or reflectance measurements provide some additional information.

There is one main disadvantage.

- The reflection from the rear surface adds intensity but no additional information to the light beam.

For reflectance measurements, the effect is often small and can be ignored. One solution for ellipsometry (in a laboratory environment) is to abrade the rear surface. Another approach is the software solution. In this case, the texture of the ellipsometric spectrum is slightly reduced. In some cases, index matching fluids can be used.

PA6.8 REFERENCES

1. Y. H. Yang and J. R. Abelson, *J. Vac. Sci. Technol. A*, **13**, 1145 (1995); F. Forcht, A. Gombert, R. Joerger, and M. Kohl, *Thin Solid Films*, **302**, 43 (1997).
2. H. G. Tompkins and P. H. Williams, *J. Vac. Sci. Technol. A*, **15**, 992 (1997).

PROTOTYPICAL ANALYSIS No. 7

Very Thick Films

PA7.1 SALIENT FEATURE

The salient feature of this prototypical analysis is the *comparison of ellipsometry and reflectometry for the characterization of very thick films.*

PA7.2 THICK PHOTORESIST ON SILICON—REFLECTOMETRY

We wish to characterize a very thick film of photoresist (~1.75 μm) on a silicon wafer. Reflectance data were acquired from this sample over the spectral range 200–800 nm at normal incidence. The reflectance spectrum is shown in Figure PA7.1.

Note that there are three distinct spectral regions in which different behavior is observed in the absolute reflectance spectrum. At the shorter wavelengths, from 200 nm to about 300 nm, the photoresist material is sufficiently absorbing to prevent light from traversing the film. As a result, no interference oscillations are observed in the measured data. At the longer wavelengths, from about 480 nm to 800 nm, the film is nonabsorbing and large-amplitude interference oscillations are observed in the reflectance data. In the intermediate spectral range from about 300 nm to 480 nm, somewhat damped interference oscillations occur, indicating that the photoresist has some optical absorption in this spectral region, but it is low enough that light can still traverse the film in this spectral range.

The simplest analysis of thick films is to choose a spectral region where the film is transparent (500–800 nm in this case), assume some value of the index of refraction of the film in this region, and fit for the thickness. If we assume the photoresist can be described by a two-term Cauchy dispersion model in this spectral region, and also assume typical values for positive resist ($n_0 = 1.55$, $n_1 = 0.005$), we obtain a film thickness of 1819.6 nm, with the fit as shown in Figure PA7.2.

The choice of seed value for the thickness determination is not obvious. Several trial-and-error attempts were made before the proper seed value was

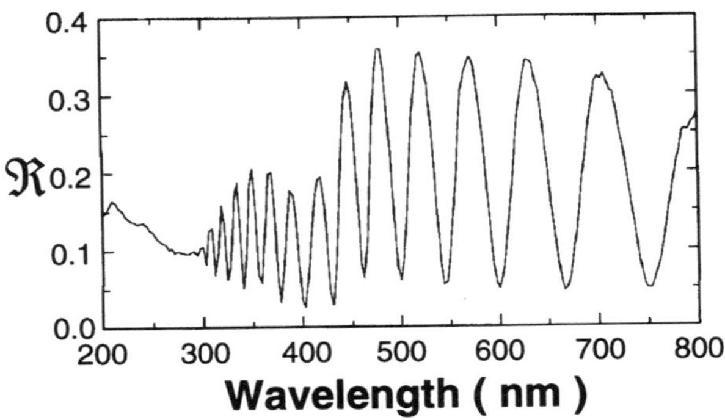

Figure PA7.1 Absolute reflectance spectrum of a thick photoresist film on silicon.

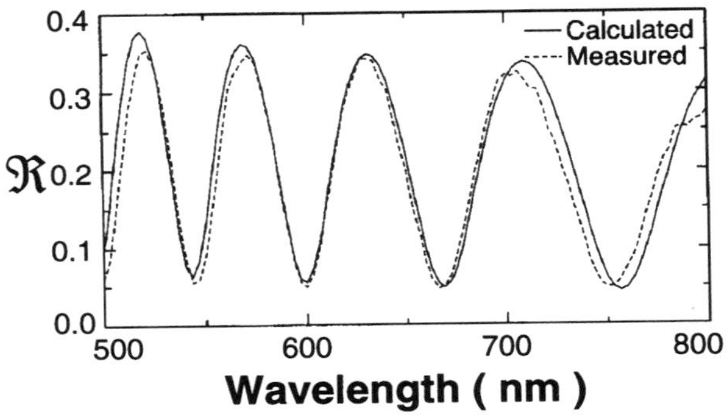

Figure PA7.2 Measured and calculated reflectance from a thick photoresist on silicon, assuming a Cauchy dispersion model (parameters listed in text).

obtained. After a trial thickness was chosen, the curve was generated and displayed. Using some of the methods discussed in Chapter 5, the next trial thickness was chosen until the thickness value was close enough to the final result for the regression algorithm to find the correct solution.

Note that neither the amplitude nor the phase of the oscillations in the measured and calculated data match well. The mismatch in amplitude is due to errors in the absolute magnitude of the index of refraction of the photoresist film, while the mismatch in phase is due to errors in the dispersion of the photoresist optical constants. If we fit for the two Cauchy parameters in addition to the thickness of the film, we obtain a film thickness of 1819.2 nm, Cauchy parameters $n_0 = 1.526$ and $n_1 = 0.0138$, and the fit shown in Figure PA7.3.

This is one of the cases where reflectance data can work very well for the measurement of the optical constants of the film. The amplitude of the interference oscillations in the reflectance data are quite sensitive to the difference in indices of refraction of the substrate, film, and ambient (air).

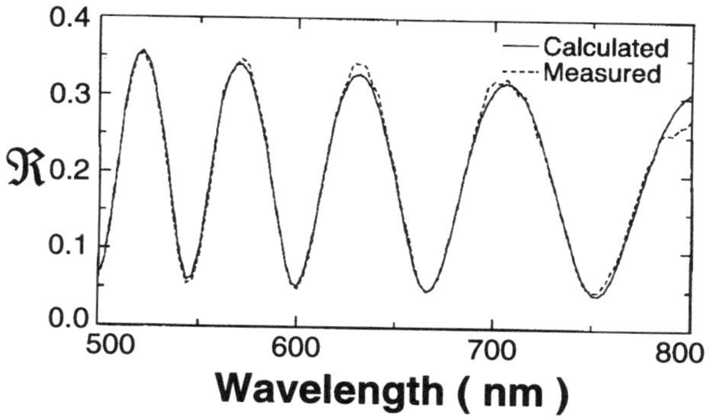

Figure PA7.3 Measured and calculated reflectance data from a thick photoresist film. Both the thickness and the Cauchy parameters for the film were varied.

PA7.3 THICK PHOTORESIST ON GLASS—ELLIPSOMETRY

The final example in this section illustrates the use of ellipsometry data for a very thick photoresist film on glass. In this case the glass optical constants were known from previous measurements, and the back surface of the glass

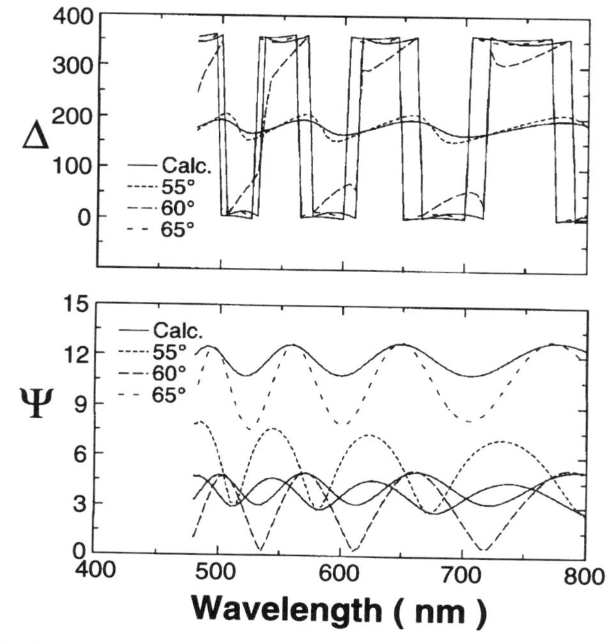

Figure PA7.4 Measured and calculated ellipsometric psi and delta from a thick photoresist film on glass. In this case, measurement instrument has a compensator, and hence delta ranges from 0° to 360°.

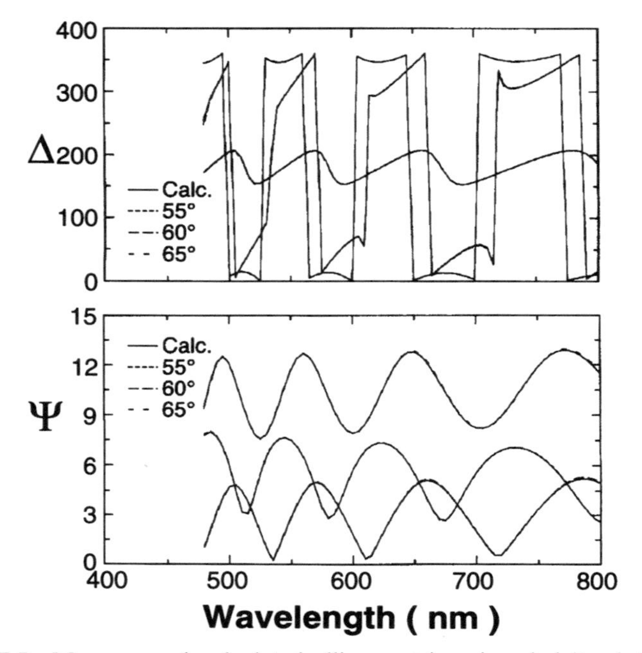

Figure PA7.5 Measure and calculated ellipsometric psi and delta data for a thick photoresist film on glass illustrated in the previous figure. In this case the index of refraction and the thickness of the film were allowed to vary.

substrate was roughened prior to measurement. We again tried the analysis by assuming a two-term Cauchy model for the resist index of refraction, with the same typical values as used previously. This yielded a film thickness of 1528.9 nm, and the fits shown in Figure PA7.4. Note that this analysis yields a final MSE of 58.

Note that while the period of the oscillations matches reasonably well, the amplitude of the oscillations in psi (in particular) do not match well at all. If we allow the Cauchy parameters to vary in addition to the thickness, we find the thickness to be 1441.8 nm, Cauchy parameters $n_0 = 1.596$, $n_1 = 0.0122$, and the final fits as shown in Figure PA7.5. Note that this analysis yields a final MSE of 1.7.

PA7.4 CONCLUSION

In conclusion, both reflectance and ellipsometric data can be used to characterize very thick films. In production applications, reflectance can be advantageous because the reflectance spectrum oscillates sinusoidally, which lends itself to simple computational algorithms for locating extrema and guessing seed values. Also, these calculations are slightly simpler owing to the reflectance being acquired at normal incidence. Finally, many production ellipsometers use a focused beam for data acquisition. This leads to a range of angles of incidence existing in the beam, which tends to smear the oscillations in the measured data.

PROTOTYPICAL ANALYSIS No. 8

Compositional Analysis of Materials

PA8.1 SALIENT FEATURE

The salient feature of this prototypical analysis is that we take advantage of our knowledge of the dependence of the optical constants of an alloy on its composition in order *to determine the composition* of an unknown sample by optical measurements. This can be done by reflectometry or ellipsometry, provided the chosen technique is sufficiently sensitive (for the given sample structure) to the optical constants of the material under study.

PA8.2 BACKGROUND

There are a number of ternary and quaternary semiconducting alloys which are of great interest to industry. Of particular importance is $Al_xGa_{1-x}As$, which is used in a large number of high-speed, optical, and telecommunications applications. The optical constants[1] of this material for several values of the aluminum concentration (x) are shown in Figure PA8.1.

Obviously, the optical constants of this alloy are very strongly dependent on the alloy concentration. As always, in order to measure the thickness of a given AlGaAs film accurately, we must know the optical constants of the film. In this case, however, we can go further by taking advantage of the strong dependence of the optical constants on the composition to determine the composition of the film as well.

We do this by constructing a simple dispersion model based on the known optical constants of the material at different concentrations. These known spectra may be obtained from the literature or from measurements on films of known concentration. We then interpolate between the known spectra to generate optical constants for values of the concentration in between known concentration values. To do this properly, the polynomials describing the dependence of the critical-point energies of the alloy as a function of the concentration must also be known. The theory behind the construction of

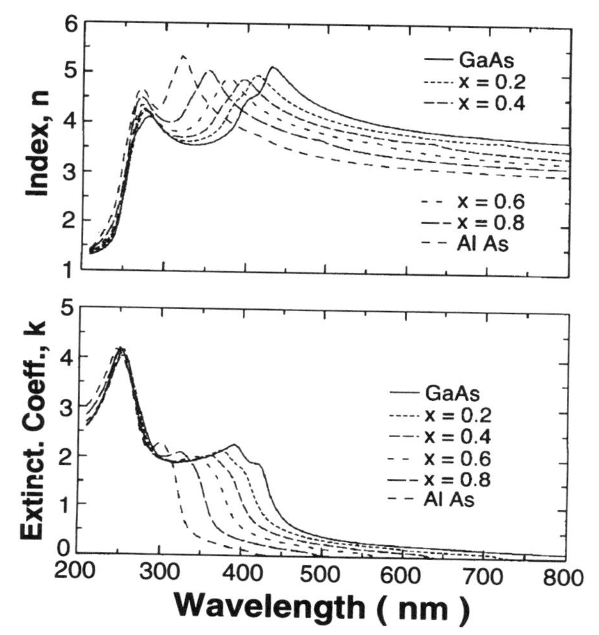

Figure PA8.1 Index of refraction and extinction coefficient of $Al_xGa_{1-x}As$ as a function of aluminimum concentration (x). Note that the endpoints are GaAs ($x = 0$) and AlAs ($x = 1$).

this model is beyond the scope of this book, so the interested reader may pursue it in the original reference.[2]

PA8.3 GaAs/AlGaAs/GaAs

This simple model parameterizes the dispersion of the optical constants of the alloy in terms of one variable parameter, the alloy concentration (x). We use this model to analyze ellipsometric data from a thick film of AlGaAs deposited on a GaAs substrate and then capped with about 50 Å GaAs. The optical model in this case consists of the GaAs substrate, the AlGaAs layer, the capping GaAs layer, and a thin surface oxide layer on top of the GaAs cap. We fit for all three layer thicknesses and the alloy fraction of the AlGaAs layer, yielding the spectra shown in Figure PA8.2.

From this analysis, the AlGaAs thickness was found to be 4654 Å, the cap layer was found to be 43 Å thick, and the surface oxide was found to be 10 Å thick. The aluminum concentration was determined to be 0.881 ± 0.004. This type of analysis can be very useful for controlling alloy concentrations, and has been applied with considerable success to the in-situ characterization and control of MBE film growth.

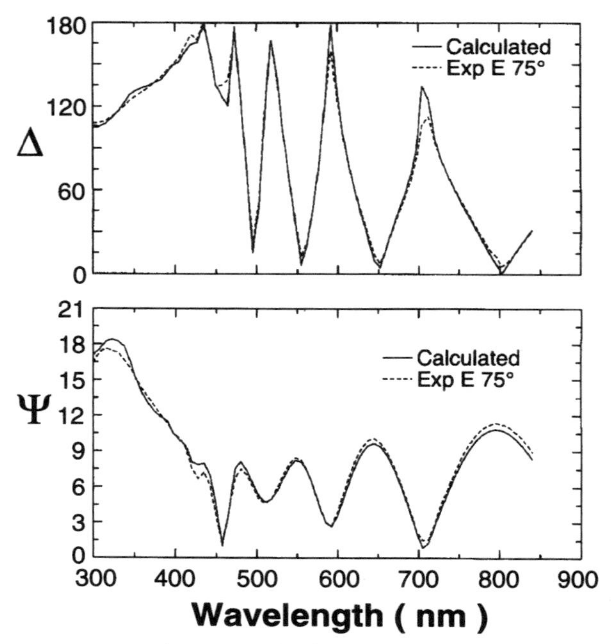

Figure PA8.2 Measured and calculated ellipsometric psi and delta data from a thick AlGaAs layer on GaAs, with a thin GaAs cap.

PA8.4 REFERENCES

1. The data for this plot were taken from software on the J. A. Woollam Co. VASE® instrument.
2. P. G. Snyder, J. A. Woollam, S. A. Alterovitz, and B. Johs, *J. Appl. Phys.*, **68**, 5925 (1990).

PROTOTYPICAL ANALYSIS No. 9

Thin Metal Films

PA9.1 SALIENT FEATURE

The salient feature of this prototypical analysis is the use of reflectance and ellipsometric data to *determine the thickness of thin metal films*. This is a very difficult problem owing to the tendency of the optical constants of metal films to vary strongly with both deposition conditions and film thickness. It is further complicated by the strong correlations generally encountered when attempting to fit for both the thickness and the optical constants of absorbing materials, and by the rather complicated dispersion of the optical constants of most metals.

PA9.2 BACKGROUND

As with any material, we can always assume some values of the optical constants of the film and fit for the thickness only, whether we are using reflectance or ellipsometry. While this technique usually yields reasonable results for dielectrics and some semiconductors, it tends to fail for metal films owing to their strong process and thickness dependence. It can be very useful, however, when applied to a single thickness of the film in production, where the process in question is under tight control. In this case, the film optical constants must be determined from a test sample, and then these optical constants can be used for production measurements. However, we still have the problem of obtaining those optical constants in the first place.

Let us assume that we wish to determine the thickness and optical constants of a given metal film. There are two general metal film structures which we wish to characterize. First, we have thin metal films deposited directly on a substrate (or another thick optically absorbing film). This is an almost impossible problem for reflectometry, and only slightly less so for ellipsometry. For any given thickness of the metal film, we can directly invert the ellipsometric data to obtain the optical constants of the metal film, obtaining a perfect fit to the data, even if multiple angles of incidence of data are acquired. This means that there are an infinite number of solutions for the thickness and optical constants of the film.

The only chance of obtaining a meaningful measurement in this case is to use a very highly constrained dispersion model for the optical constants of the metal, so that we are fitting for the thickness of the metal and a few parameters in the dispersion model. Lorentz oscillator models are most commonly used in this case. Even then, we must be very careful to watch the correlation between the dispersion model parameters and the thickness, and there is always a strong chance that the solution obtained is not correct.

The second structure is a metal film on top of a transparent film on an absorbing substrate. Under certain circumstances, this structure can be analyzed and we give one example.

One of the strongest methods of determining optical constants of any material is to have several different films of the same material, differing only in thickness. This is true for metal films as well as dielectric films, and we give an example of this method.

PA9.3 THIN TITANIUM ON SILICON

Let us consider ellipsometric data from a thin (~250 Å) titanium film deposited directly on silicon. We first try to fit the data using optical constants for titanium taken from the literature[1] (for bulk titanium), yielding a thickness of 318 Å and the fits shown in Figure PA9.1.

Next, we try to improve this fit by using a five-oscillator Lorentz model for the optical constants of the titanium film. The starting values of the oscillator parameters were determined by fitting the Lorentz oscillator model to the literature optical constants of titanium. We then experiment with varying different oscillator parameters to find which parameters yield the best fit to the data without introducing unacceptable parameter correlations. This can be a lengthy process, and does not lend itself to production-worthy algorithms. The final result was obtained by varying the amplitudes of the five oscillators, yielding a thickness of 243 Å and the results shown in Figure PA9.2. The resulting optical constants are shown in Figure PA9.3 along with the optical constants from the literature[1] for comparison.

PA9.4 THIN TITANIUM NITRIDE ON OXIDE ON SILICON

The most accurate analysis of thin metal films is obtained by performing ellipsometric analysis on the second type of structure of interest, a thin metal film on a thick dielectric film on a substrate.[2] One of the authors (WAM) showed that acquiring ellipsometric spectra over a reasonably wide range of angles of incidence allows the unique determination of the thicknesses of the

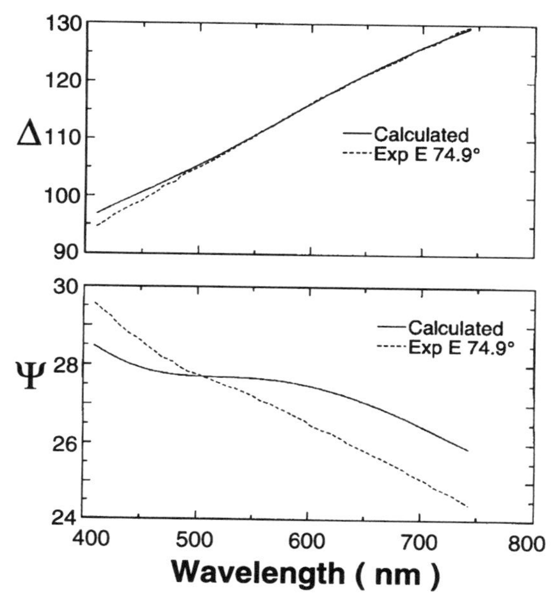

Figure PA9.1 Measured and calculated ellipsometric psi and delta data for a thin film of titanium on silicon. Literature optical constants[1] were used for the titanium film.

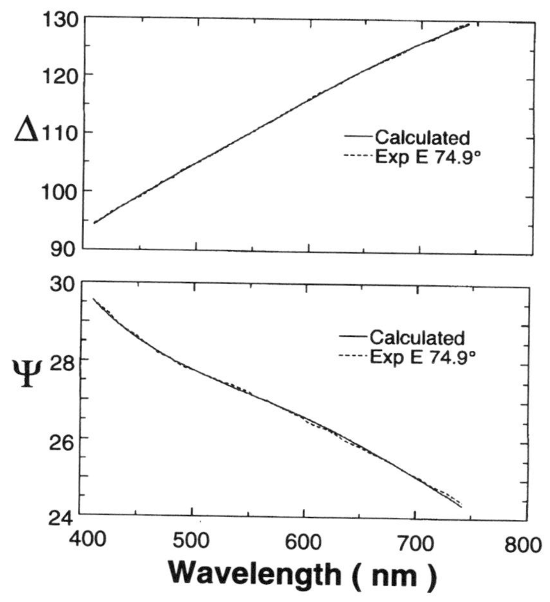

Figure PA9.2 Measured and calculated ellipsometric data for thin titanium film on silicon, varying the thickness and five Lorentz oscillator amplitudes.

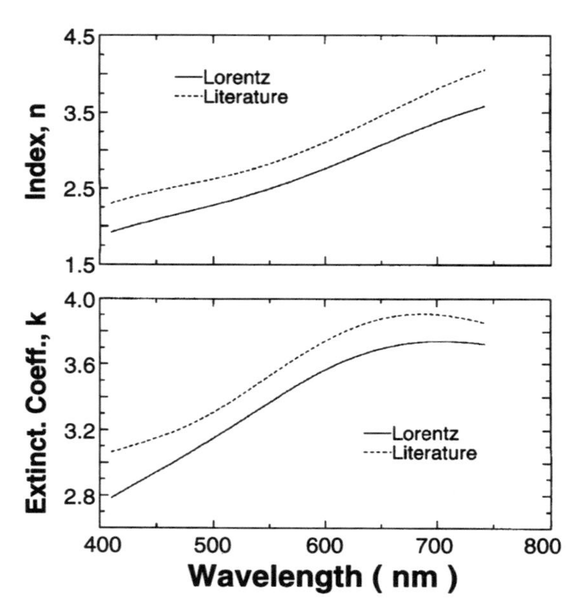

Figure PA9.3 The optical constants obtained from the analysis of the data shown in Figure PA9.2, along with those from the literature[1] for comparison.

metal and dielectric films, as well as the optical constants of the metal at each wavelength, without requiring the use of a dispersion model. Figure PA9.4 shows the results of such an analysis of a thin TiN film deposited on a thick thermal silicon dioxide film on silicon. The TiN film was found to be 240 Å thick and the oxide layer was 5055 Å thick. Note that the inclusion of a 20 Å surface oxide on the TiN film (modeled as titanium dioxide) reduced the final MSE from 4.8 to 2.1 and eliminated oscillatory artifacts in the resulting TiN optical constant spectra.

This is a very powerful method for directly determining the optical constants of thin metal films, particularly since it can be performed from ellipsometric data alone. Another powerful means of performing this analysis is to deposit the metal film on a glass substrate and simultaneously analyze ellipsometric and transmittance data from the film; however, the transmittance measurement is generally less accurate than ellipsometry.

PA9.5 USING MULTIPLE FILMS OF THIN METAL

For this example, we develop the optical constants of thin chromium. The fabricator was asked to make several samples of thin chromium on silicon, with the only difference being the thickness. The resulting ellipsometric data

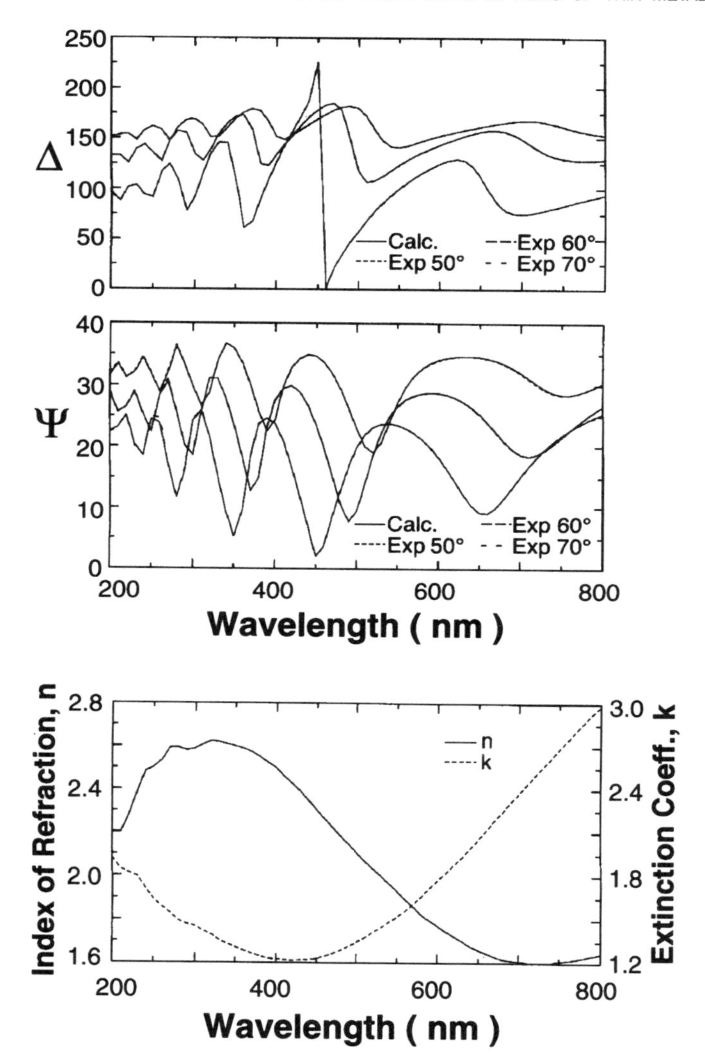

Figure PA9.4 Measured and calculated ellipsometric spectra from a thin TiN film on a thick silicon dioxide film on silicon. The optical constants obtained for the TiN film are also shown.

are shown in Figure PA9.5 as 1, 2, 3, and 4, with film 1 expected to be the thinnest and 4 expected to be the thickest, based on deposition parameters.

If the literature values[3] of the optical constants for chromium were used as received, the resulting thickness values were 20, 44, 62, and 90 Å, but the MSE was excessively large (MSE = 227).

Using these literature optical constants and these thickness values for seed values, all four films were then analyzed simultaneously, with the regression

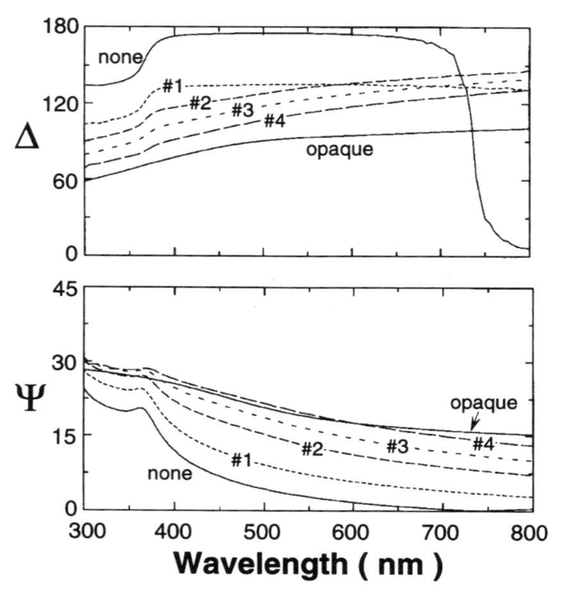

Figure PA9.5 Delta/psi data for four samples of thin chromium on silicon. Samples #1 is expected to be the thinnest and #4 is expected to be the thickest, based on deposition parameters. Also shown are the delta/psi values for no film and for a film which is thick enough to be opaque. This is calculated from the optical constants which were obtained.

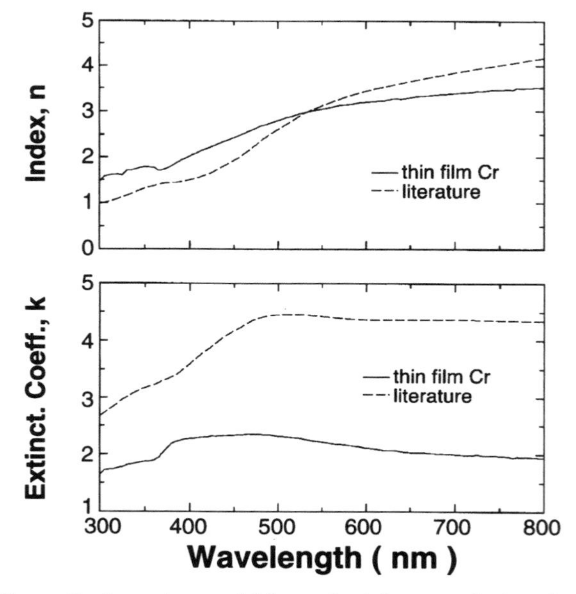

Figure PA9.6 The optical constants which resulted from analyzing the data shown in Figure PA9.5. Also shown are the literature values[3] for chromium.

parameters being n and k for each wavelength (independently) and the four thicknesses. The resulting thickness values were 42, 90, 137, and 205 Å, and the resulting optical constants are shown in Figure PA9.6 as thin film Cr along with the literature values. The MSE was 36.

We can now calculate the delta/psi values for an opaque film with these optical constants, and these are also shown in Figure PA9.5. Also shown in this figure are the delta/psi values for no film (i.e., film-free silicon). This shows the gradual progression of the SE spectrum as the film thickness increases. It also shows that differences of tens of angstroms can be distinguished.

Related work for Cu, Co, Ta, NiFe, and FeMn is described by Tompkins et al.[4] In addition to considering single layers, that work showed that in some cases double layers could be distinguished from each other.

PA9.6 COMMENTS AND SUMMARY

A thin metal film on a semiconductor or dielectric substrate changes the optical properties significantly. In principle, both spectroscopic ellipsometry and reflectometry can be used to determine thickness values of the thin metal layers. However, the optical constants of a thin metal layer are almost invariably different from those of a bulk sample of the same material, and it is essential to determine the optical constants of the material in the thin-layer form. We have suggested several approaches for this determination using spectroscopic ellipsometry. One has to recognize, however, that none of these methods are particularly straightforward, and the optical constant determination would probably have to be done on an analytical instrument rather than a metrology instrument.

PA9.7 REFERENCES

1. Literature values for optical constants for titanium were taken from the software on the J. A. Woollam Co. VASE® instrument.
2. W. A. McGahan, B. Johs, and J. A. Woollam, *Thin Solid Films*, **234**, 443 (1993).
3. The literature values for the optical constants of chromium were taken from D. W. Lynch and W. R. Hunter, in *Handbook of Optical Constants of Solids II*, edited by E. D. Palik, Academic Press, New York, 1991.
4. H. G. Tompkins, T. Zhu, and E. Chen, *J. Vac. Sci. Technol. A*, **16**, 1297 (1998).

PROTOTYPICAL ANALYSIS No. 10

Photoresist Optical Constants

PA10.1 SALIENT FEATURE

The salient feature of this prototypical analysis is the use of ellipsometric data to directly *determine the optical constants of an organic material.* This will illustrate one of the most powerful techniques in existence for the analysis of ellipsometric data from unknown films—*the use of a "window" in the spectrum where the film is transparent* to determine the thickness of the film. Note that photoresist is also discussed in prototypical analyses No. 3 and No. 7.

PA10.2 BACKGROUND

As shown in prototypical analysis No. 9, it is very difficult to determine the optical constants of an absorbing film directly. To do so generally requires prior knowledge of a good dispersion model for the material under study. Photoresists and organic films in general can exhibit considerable structure in their optical constants, particularly in the UV portion of the spectrum, and it is impossible to construct a single general dispersion model which can handle them all. However, we can take advantage of their transparency in the visible and NIR portions of the spectrum to perform an exact determination of their optical constants at all wavelengths. We first determine a spectral range over which the film is transparent, and fit the data in this region to determine the thickness and index of refraction (via a Cauchy dispersion model) of the film. Then, we fix the thickness of the film and invert the ellipsometric data at each measured wavelength to obtain the optical constants of the film. This technique works well for any material which is transparent in some part of the measured spectral range.

PA10.3 PHOTORESIST ON SILICON

Ellipsometric data were acquired for a photoresist film on silicon. To determine the spectral range over which the film is transparent, we model the film with a two-term Cauchy dispersion model and fit for both Cauchy parameters and the thickness of the film. We then incrementally shrink the spectral range being analyzed and try the fit again. When reducing the spectral range does not significantly improve the fit, we have found a range where the film is transparent. In this case, fitting the data from 400 to 800 nm yields an MSE of 202, fitting from 500 to 800 nm yields an MSE of 47, and fitting from 550 to 800 nm yields an MSE of 9. Fitting from 650 to 800 nm yields an MSE of 3.4, and reducing the spectral range further does not significantly improve the MSE. This analysis yields a film thickness of 1823 Å, Cauchy parameters $n_0 = 1.680$, $n_1 = 0.0061$, and $n_2 = 0.0069$, and the fits shown in Figure PA10.1.

We then fix the film thickness at 1823 Å and fit directly for the optical constants at each measured wavelength. The results are shown in Figure PA10.2.

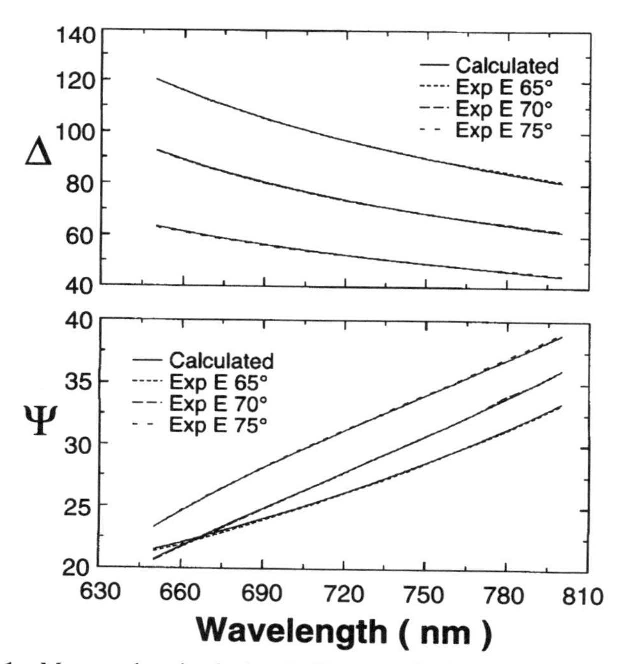

Figure PA10.1 Measured and calculated ellipsometric data from a photoresist film on silicon.

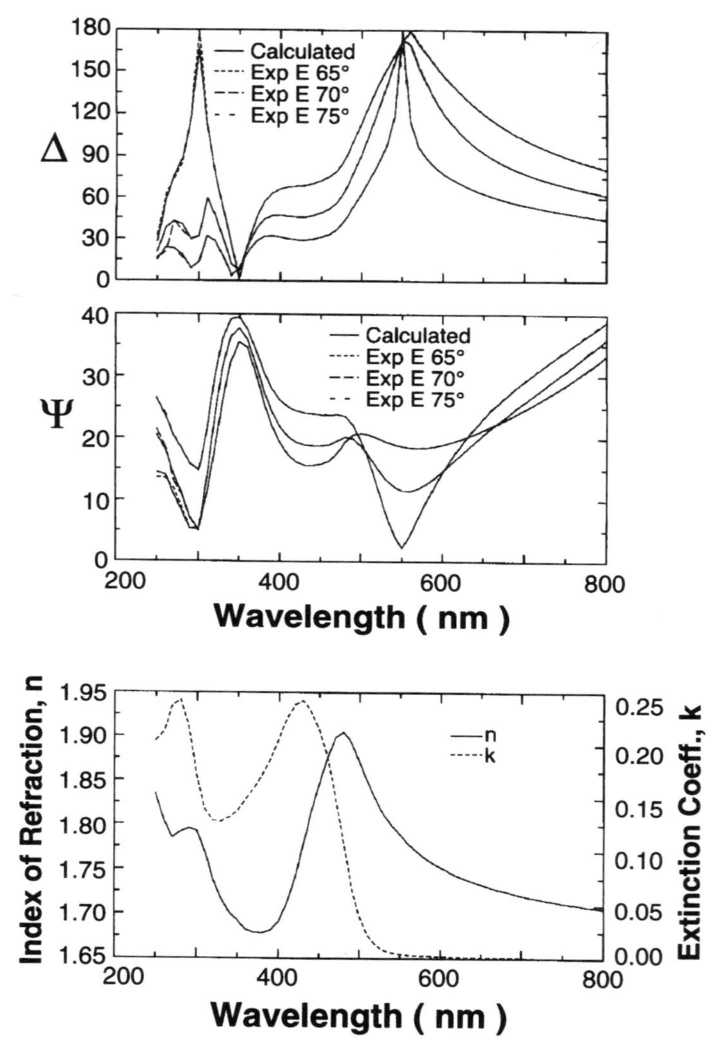

Figure PA10.2 Measured and calculated ellipsometric data from point-to-point fit for the optical constants of the photoresist film, with the thickness fixed at the value obtained from the Cauchy fit. The optical constants are also shown.

PA10.4 SUMMARY

This procedure works very well provided the film is truly transparent in the spectral region used for the Cauchy fit. If this is not the case, the extinction coefficient generated by the point-by-point fit in the region of the spectrum in which the original Cauchy fit was performed will not be zero.

This is a very powerful procedure for the determination of the optical constant spectra of organic films which have complicated dispersion in the UV spectral range, and can also be applied to inorganic films (oxynitrides, for example) provided they are transparent somewhere in the measured spectral range.

APPENDICES

APPENDIX A

Regression Algorithms

A.1 THE MERIT FUNCTION

In Chapter 9, we defined the merit function[1,2] which is used to judge the quality of the match between the measured and calculated data as the mean-squared error (MSE):

$$\text{MSE} = \frac{1}{N - M} \sum_{1}^{N} \left(\frac{y_i - y(\bar{x}, \bar{a})}{\sigma_i} \right)^2 \tag{A.1}$$

In this equation, y_i denote the experimentally measured data values, $y(\bar{x}, \bar{a})$ are the calculated model data points, and σ_i is the standard deviation of the ith data point. The vector $\bar{x}$ contains all known parameters in the model, while the vector $\bar{a}$ contains the model parameters which are to be varied in order to fit the experimental data. There are a total of N data points, and a total of M variable parameters. This function will exhibit a minimum value of zero when the calculated model data exactly matches the experimentally measured data.

A.2 ANALYSIS PROCEDURE

The analysis procedure is:

- to collect a set of measured experimental data points;
- to propose a model function which can be used to calculate data points corresponding to the measured ones;
- to choose a set of variable parameters in the model which we want to adjust so that the calculated data match the experimental data as closely as possible;
- to use the regression procedure to reduce the MSE to a minimum value.

The merit function has been chosen. The next step is to define an algorithm which can be used to adjust the values of the variable parameters in the model. The goal of this algorithm is to get the calculated model data to match the experimental data as closely as possible, giving a global minimum for the MSE value. There are a number of algorithms in the literature to perform this minimization, but only a few of the most popular ones will be discussed here.

A.3 ALGORITHMS

The two simplest algorithms[2] for minimizing the MSE are the gradient method (also called the steepest descent method) and the inverse Hessian method. Both of these methods rely on the information obtained by taking derivatives of the MSE with respect to the variable parameters in the optical model. The gradient method relies on the fact that the derivative of the MSE with respect to the model parameters (the gradient of the MSE surface) will be nonzero away from the minimum, while the inverse Hessian method takes advantage of the fact that the gradient of the MSE with respect to the variable parameters is zero at the minimum.

The gradient method is probably the simplest method to implement, but can require a lot of computational time to find the minimum. In this method, we calculate the derivative of the MSE with respect to each model parameter. This tells us whether we need to increase or decrease the given parameter in order to decrease the MSE, and also tells us how steep the MSE surface is with respect to the given parameter (i.e., what the slope of the MSE surface is). The gradient vector is calculated as follows:

$$\frac{\partial \text{MSE}}{\partial a_k} = -2\left[\frac{y_i - y(\bar{x}, \bar{a})}{\sigma_i} \cdot \frac{\partial y(\bar{x}, \bar{a})}{\partial a_k}\right] \tag{A.2}$$

In many cases this derivative is evaluated numerically by adjusting the variable parameter slightly and using Newton's approximation to evaluate the derivative. Note that the convergence rate of the fitting algorithm can be rather sensitive to the parameter increment used to evaluate the derivative, and many analysis programs include this increment as a user-defined parameter.

The gradient vector shown above tells us how steep the MSE surface is with respect to the variable parameters in the model and, more importantly, which direction is "down." Given this information, we can always adjust the variable parameters so that we take a step along the direction of the gradient vector in the downward direction. Such a step should always reduce the MSE. When we can no longer reduce the MSE in this manner, we should be at the minimum.

There are several drawbacks to this approach. First, we have no idea how large or small a step to take along the gradient. If the steps are too small, the algorithm will require an excessively large number of iterations to converge. If they are too large, the algorithm may overshoot the minimum. Second, there are often many local minima of the MSE function, and the gradient method has a tendency to find such minima and become trapped. All in all, the gradient method tends to work well as a last resort if the current variable parameter values are far from the best-fit values, but can be problematic close to the final solution.

The second common regression algorithm is the inverse Hessian method. This method is based on the fact that the MSE function is a sum of squares, and that the gradient of the MSE with respect to the variable model parameters will be zero at the MSE minimum. Given these assumptions, it is possible to calculate the exact location of the MSE minimum analytically from the current values of the model parameters and the first and second derivatives of the MSE with respect to the variable model parameters. This method works extremely well if the current parameter values are very close to the final best-fit solution, but has a tendency to require a lot of iterations if the current variable parameters are very far from the true best-fit solution.

A.4 THE COMBINATION

Note that the gradient and inverse Hessian methods are somewhat complementary. The inverse Hessian method works well close to the minimum, while the gradient method is better far from the minimum. The best solution is obviously a hybrid of the two algorithms. This hybrid was developed by Marquardt, based on earlier suggestions by Levenberg, and is an elegant merger of the gradient and inverse Hessian algorithms. The algorithm, referred to as the Levenberg–Marquardt (LM) algorithm, is developed in full mathematical detail by Press et al.[2] and we will not present those details here. In short, a single parameter λ (the Marquardt parameter, not to be confused with wavelength) is used to vary smoothly between the gradient and inverse Hessian algorithms. When λ is very large, the algorithm behaves like the gradient method. When λ is very small, the algorithm behaves like the inverse Hessian algorithm. When λ is intermediate in value, the algorithm is a blend of the two.

Most LM algorithms are implemented by initially assuming a small value for λ (i.e., assuming the initial parameter guesses are close to their best-fit values) and calculating a step. If the step improves the MSE, λ is reduced by a factor of 10 (typically) and another iteration is performed. If the step does not improve the MSE, the variable parameters are restored to the values they had before the step and λ is increased, typically by a factor of 10. Iteration usually continues until the MSE cannot be reduced any further or until a maximum number of iterations has been reached.

The behavior of an LM algorithm during a given data analysis procedure can be a very useful diagnostic tool. In particular, the number of iterations required to reach the best-fit solution and the amount by which the MSE is reduced at each iteration can be indicative of problems in the optical model and/or the measured data. If the optical model for the sample under test is physically accurate, the initial guesses at the variable model parameters are reasonably close, and there are no strong correlations among the variable model parameters, the LM algorithm will usually converge to the best-fit solution in a few iterations. Parameter correlations have the strongest effect on the convergence rate of the LM algorithm. If there are very strong correlations among the variable model parameters, the LM algorithm tends to require a very large number of iterations to reach the best-fit minimum, and each iteration tends to reduce the MSE by a very small amount.

A.5 REFERENCES

1. G. E. Jellison, Jr., *Appl. Opt.*, **30**, 3354 (1991).
2. W. H. Press, B. P. Flannery, S. A. Teukolsky, and W. T. Vetterling, *Numerical Recipes*, Cambridge University Press, Cambridge, 1988.

APPENDIX B

Maxwell's Equations and the Wave Equation

B.1 INTRODUCTION

The description of light as an electromagnetic wave developed in a rather fragmentary way until James Clerk Maxwell (1831–1879) provided a unifying description in a paper presented before the Royal Society in 1864. The paper was entitled "A Dynamical Theory of the Electromagnetic Field." In this paper, Maxwell proposed a theory which required the vibrations to be strictly transverse and provided a definite connection between light and electricity.[1] The results of this theory were expressed as four equations which are known as *Maxwell's equations*.

For perspective, consider that for a mechanical motion physics problem we first write the differential equations which describe the forces on the mechanical body. We then put in the boundary conditions, and solve for an equation which describes the trajectory of the mechanical body as a function of time and/or position. For an electromagnetic physics problem, the Maxwell's equations are analogous to the differential equation, and we shall derive the *wave equation* which describes the behavior of the entity which is waving as a function of time and position. At one point in history, this was considered to be the "aether" particles. After the development of the electromagnetic theory, the entity which is waving is considered to be the amplitude of the electric vector (and, correspondingly, the amplitude of the magnetic vector).

We list the four equations below, first in vector calculus notation and later in differential equation form.[2] We will describe the various quantities in a heuristic manner, to help convey the basic ideas.

Maxwell's equations, in vector notation (there is a review of vector calculus notation at the end of this appendix) in their most general form (in gaussian units), are:

$$\nabla \times \bar{H} - \frac{1}{c}\frac{\partial \bar{D}}{\partial t} = \frac{4\pi}{c}\bar{j} \tag{B.1}$$

$$\nabla \times \bar{E} + \frac{1}{c}\frac{\partial \bar{B}}{\partial t} = 0 \tag{B.2}$$

$$\nabla \cdot \bar{D} = 4\pi\rho \tag{B.3}$$

$$\nabla \cdot \bar{B} = 0 \tag{B.4}$$

The first of these equations is Ampere's law dealing with how currents induce magnetic fields. The second is Faraday's law dealing with how changing magnetic fields induce electric fields. The third equation indicates that lines of electrical force emanate from charges and the fourth equation indicates that there are no free magnetic poles, that is, no "magnetic charge" analogous to the electric charge in Eq. B.3.

The electromagnetic field is represented by the basic field vectors $\bar{E}$ and $\bar{B}$, which are the electric vector and magnetic induction, respectively. However, these two quantities are not sufficient to describe the interaction of light with matter. We introduce three other vectors to describe the effect of the basic field vectors on matter. These are the electric displacement $\bar{D}$, the magnetic field vector $\bar{H}$, and the electric current density $\bar{j}$. ρ is the electric charge density. The vectors describing the interaction with matter are related to the basic field vectors by

$$\bar{j} = \sigma\bar{E} \tag{B.5}$$

$$\bar{D} = \varepsilon\bar{E} \tag{B.6}$$

$$\mu\bar{H} = \bar{B} \tag{B.7}$$

where σ is the specific conductivity, ε is the dielectric constant, and μ is the magnetic permeability. Note that Eq. B.5 is a form of Ohm's law. Materials for which σ is negligibly small are called insulators or dielectrics. For most materials, μ is practically unity. If μ is significantly different from unity, either greater or less, the material is magnetic. In optics, we are concerned primarily with ε, and in the case of conductors, with σ. In what follows, we consider only materials where no free charge exists, and hence $\rho = 0$. The quantities σ, ε, and μ are properties of the material through which the light passes.

Using these relationships, we rewrite Maxwell's equations as

$$\nabla \times \bar{H} - \frac{\varepsilon}{c}\frac{\partial \bar{E}}{\partial t} = \frac{4\pi\sigma}{c}\bar{E} \tag{B.8}$$

$$\nabla \times \bar{E} + \frac{\mu}{c}\frac{\partial \bar{H}}{\partial t} = 0 \tag{B.9}$$

$$\nabla \cdot \bar{E} = 0 \tag{B.10}$$

$$\nabla \cdot \bar{H} = 0 \tag{B.11}$$

B.2 DIFFERENTIAL EQUATION FORM OF MAXWELL'S EQUATIONS

To illustrate the differential equation form, we will, for simplicity, consider a dielectric medium, that is, $\sigma = 0$. The two *curl* equations must be shown in their component parts. Equation B.8 becomes

$$\begin{aligned}
\frac{\partial H_z}{\partial y} - \frac{\partial H_y}{\partial z} &= \frac{\varepsilon}{c}\frac{\partial E_x}{\partial t} \\
\frac{\partial H_x}{\partial z} - \frac{\partial H_z}{\partial x} &= \frac{\varepsilon}{c}\frac{\partial E_y}{\partial t} \\
\frac{\partial H_y}{\partial x} - \frac{\partial H_x}{\partial y} &= \frac{\varepsilon}{c}\frac{\partial E_z}{\partial t}
\end{aligned} \tag{B.12}$$

Equation B.9 becomes

$$\begin{aligned}
\frac{\partial E_z}{\partial y} - \frac{\partial E_y}{\partial z} &= -\frac{\mu}{c}\frac{\partial H_x}{\partial t} \\
\frac{\partial E_x}{\partial z} - \frac{\partial E_z}{\partial x} &= -\frac{\mu}{c}\frac{\partial H_y}{\partial t} \\
\frac{\partial E_y}{\partial x} - \frac{\partial E_x}{\partial y} &= -\frac{\mu}{c}\frac{\partial H_z}{\partial t}
\end{aligned} \tag{B.13}$$

and Eqs. B.10 and B.11 become

$$\frac{\partial E_x}{\partial x} + \frac{\partial E_y}{\partial y} + \frac{\partial E_z}{\partial z} = 0 \tag{B.14}$$

and

$$\frac{\partial H_x}{\partial x} + \frac{\partial H_y}{\partial y} + \frac{\partial H_z}{\partial z} = 0 \tag{B.15}$$

respectively. These are Maxwell's equations (for a dielectric medium with no excess charge) in differential equation form.

B.3 THE WAVE EQUATION

B.3.1 Dielectric Media

To convert Maxwell's equations into a description of the light wave, let us again restrict ourselves to a dielectric medium ($\sigma = 0$) and later generalize to include conductors. In vector notation, for dielectric media with no free charge,

Maxwell's equations are

$$\nabla \times \bar{H} - \frac{\varepsilon}{c}\frac{\partial \bar{E}}{\partial t} = 0 \tag{B.16}$$

$$\nabla \times \bar{E} + \frac{\mu}{c}\frac{\partial \bar{H}}{\partial t} = 0 \tag{B.17}$$

$$\nabla \cdot \bar{E} = 0 \tag{B.18}$$

$$\nabla \cdot \bar{H} = 0 \tag{B.19}$$

Let us reduce these to two equations. Taking the curl of the first equation, we obtain

$$\nabla \times (\nabla \times \bar{H}) - \nabla \times \left(\frac{\varepsilon}{c}\frac{\partial \bar{E}}{\partial t}\right) = 0$$

Using the identity

$$\nabla \times (\nabla \times \bar{Q}) \equiv \nabla(\nabla \cdot \bar{Q}) - \nabla^2 \bar{Q} \tag{B.20}$$

and transforming the time derivative, we have

$$\nabla(\nabla \cdot \bar{H}) - \nabla^2 \bar{H} - \frac{\varepsilon}{c}\frac{\partial(\nabla \times \bar{E})}{\partial t} = 0$$

Substituting from Eqs. B.19 and B.17 for $\nabla \cdot \bar{H}$ and $\nabla \times \bar{E}$, we obtain

$$\nabla^2 \bar{H} = \frac{\mu\varepsilon}{c^2}\frac{\partial^2 \bar{H}}{\partial t^2} \tag{B.21}$$

and in a similar manner,

$$\nabla^2 \bar{E} = \frac{\mu\varepsilon}{c^2}\frac{\partial^2 \bar{E}}{\partial t^2} \tag{B.22}$$

These are standard equations of wave motion and suggest the existence of electromagnetic waves propagated with a velocity of

$$v = \frac{c}{\sqrt{\mu\varepsilon}} \tag{B.23}$$

Recall that the index of refraction is defined as $n = c/v$, and hence

$$n = \sqrt{\mu\varepsilon} \tag{B.24}$$

That Eqs. B.21 and B.22 have the same form suggests that the magnetic and electric vectors will be in phase. It can be shown[1–3] that the electric and magnetic vectors are perpendicular to each other and to the direction of propagation of the wave. The solutions to Eqs. B.21 and B.22 consist of sines, cosines, and/or imaginary exponentials.

Suppose that ν (nu) is the frequency of oscillations of the wave in cycles per second, v is the velocity, and λ is the wavelength. Then

$$\nu = \frac{v}{\lambda} \tag{B.25}$$

Note that when a light wave passes from one medium to another, both the velocity v and the wavelength λ will change in proportion, but the frequency ν will remain the same. We define the angular frequency ω, in radians per second, as

$$\omega = 2\pi\nu = 2\pi\frac{v}{\lambda} \tag{B.26}$$

Note that λ is the wavelength in the material of interest. This is related to the vacuum wavelength λ_0 by the relation $\lambda_0 = n\lambda$. Stated explicitly, ω contains information about the wave, but not about the material through which it travels. ε, and subsequently v, and λ contains information about the material. Let us take the time dependence of the electric and magnetic fields to be proportional to $e^{j\omega t}$ so that $\bar{E}(x, y, z, t) = \bar{E}_0(x, y, z)e^{j\omega t}$, which leads to

$$\frac{\partial \bar{H}}{\partial t} = j\omega\bar{H} \qquad \text{and} \qquad \frac{\partial \bar{E}}{\partial t} = j\omega\bar{E} \tag{B.27}$$

Equation B.22 becomes

$$\nabla^2 \bar{E}_0 = -\frac{\mu\varepsilon}{c^2}\,\omega^2 \bar{E}_0 \tag{B.28}$$

and with Eqs. B.23 and B.26, this becomes

$$\nabla^2 \bar{E}_0 = -\frac{(2\pi)^2}{\lambda^2}\,\bar{E}_0 \tag{B.29}$$

The solution to the space part of the equation might be written in vector format as

$$\bar{E}_0 = \bar{E}^0 e^{-j\omega(\sqrt{\mu\varepsilon}/c)\bar{r}\cdot\bar{s}}$$

or

$$\bar{E}_0 = \bar{E}^0 e^{-j(2\pi/\lambda)\bar{r}\cdot\bar{s}}$$

where $\bar{r}$ is the position vector, $\bar{s}$ is a unit vector in the direction of propagation, and $\bar{E}^0$ is the amplitude. To reiterate, $\bar{E}$ is a vector-valued function of position and time, $\bar{E}_0$ is a vector-valued function of position, and $\bar{E}^0$ is a vector constant. The complete solution to Eq. B.22 might be

$$\bar{E}(\bar{r}, t) = \bar{E}^0 e^{-\mathrm{j}(2\pi/\lambda)\bar{r}\cdot\bar{s}} e^{\mathrm{j}\omega t} \tag{B.30}$$

In a single dimension, a solution might be

$$E_z = E_z^0 \sin\left(-\frac{2\pi}{\lambda}x + \omega t\right) \tag{B.31}$$

Some other useful forms of Eq. B.30 are

$$\bar{E}(\bar{r}, t) = \bar{E}^0 e^{\mathrm{j}\omega(t - \bar{r}\cdot\bar{s}/v)} \tag{B.32}$$

and

$$\bar{E}(\bar{r}, t) = \bar{E}^0 e^{-\mathrm{j}(2\pi/\lambda)(\bar{r}\cdot\bar{s} - vt)} \tag{B.33}$$

which are simple algebraic manipulations using Eq. B.26.

B.3.2 INCLUDING CONDUCTORS

The above treatment deals with the wave equation for $\sigma = 0$. When this is not the case, the situation is a bit more involved.[4] Again, let us assume that the time dependence of the electric and magnetic fields is proportional to $e^{\mathrm{j}\omega t}$. Equations B.8 and B.9 become

$$\nabla \times \bar{H}_0 - \frac{\varepsilon}{c}\mathrm{j}\omega\bar{E}_0 = \frac{4\pi\sigma}{c}\bar{E}_0$$

or

$$\nabla \times \bar{H}_0 = \left(\frac{\varepsilon}{c}\mathrm{j}\omega + \frac{4\pi\sigma}{c}\right)\bar{E}_0 \tag{B.34}$$

and

$$\nabla \times E_0 + \frac{\mu}{c}\mathrm{j}\omega\bar{H}_0 = 0 \tag{B.35}$$

As before, taking the curl of Eq. B.35, using the identity in Eq. B.20, and substituting for $\nabla \cdot \bar{E}$ from Eq. B.10 and $\nabla \times \bar{H}$ from Eq. B.34, we obtain

$$\nabla^2\bar{E}_0 + \frac{\mu}{c^2}\left(\varepsilon - \mathrm{j}\frac{4\pi\sigma}{\omega}\right)\omega^2\bar{E}_0 = 0 \tag{B.36}$$

Note the similarity of this equation to Eq. B.28. The ε term in that equation has been replaced by a complex number

$$\tilde{\varepsilon} = \varepsilon - \mathrm{j}\,\frac{4\pi\sigma}{\omega} \tag{B.37}$$

or

$$\tilde{\varepsilon} = \varepsilon_1 - \mathrm{j}\varepsilon_2$$

The net effect of the imaginary part of this term is that the space part of the exponent in Eq. B.30 is no longer pure imaginary, and that there is now a part which causes *attenuation as a function of distance.*

The complex index of refraction $\tilde{N}$ is defined, analogous to Eq. B.24, as

$$\tilde{N} = \sqrt{\mu\tilde{\varepsilon}} \tag{B.38}$$

Thus the real and imaginary parts are usually separated out, and this is written as

$$\tilde{N} = n - \mathrm{j}k \tag{B.39}$$

where n is the index of refraction (which leads to confusion) and k is the extinction coefficient.

B.3.3 The Magnetic Field Wave

It is generally accepted that the human eye senses the electric field, and the corresponding magnetic field variation is usually ignored. It is important for the derivation of some equations, however, to consider the magnetic field as well. Equations B.30–B.34 deal with the electric vector and were derived from Eqs. B.17–B.20. The magnetic wave equation can also be derived from the same original equations and is similar in nature. It has an amplitude $\bar{H}^0$ and the same exponential term as the electric wave equation. This is to say that the oscillations of the electric and magnetic fields are in phase.

Direction

We can readily show that the vector $\bar{H}$ is perpendicular to the vector $\bar{E}$. By combining Eqs. B.16 and B.17 with Eqs. B.27, we obtain

$$\nabla \times \bar{H} = \frac{\varepsilon}{c}\,\mathrm{j}\omega\bar{E} \tag{B.40}$$

and

$$\nabla \times \bar{E} = -\frac{\mu}{c}\,\mathrm{j}\omega\bar{H} \tag{B.41}$$

Since the vector $\nabla \times \bar{H}$ is perpendicular to the vector $\bar{H}$, Eq. B.40 implies that the vector $\bar{E}$ is perpendicular to the vector $\bar{H}$. Although not shown here, Eqs. B.40 and B.41 can also be used to show that both $\bar{E}$ and $\bar{H}$ are mutually perpendicular to the direction of propagation, given by the unit vector $\bar{s}$, used in Eq. B.30.

Magnitude

The relationship of the magnitude of the vector $\bar{E}$ to the magnitude of the vector $\bar{H}$ can be illustrated by supposing, without loss of generality, that the wave is propagating in the positive x direction and that the electric vector is vibrating in the z direction.

Under these conditions,

$$\bar{r} \cdot \bar{s} = x \tag{B.42}$$

and using Eq. B.32,

$$E_x = E_z = 0 \qquad \text{and} \qquad E_y = E_y^0 \, \mathrm{e}^{\mathrm{j}\omega(t - x/v)} \tag{B.43}$$

and

$$H_x = H_y = 0 \qquad \text{and} \qquad H_z = H_z^0 \, \mathrm{e}^{\mathrm{j}\omega(t - x/v)} \tag{B.44}$$

We now use Maxwell's equations in their component form (Eqs. B.12 and B.13). The only nonzero partial derivatives are

$$\frac{\partial E_y}{\partial t}, \frac{\partial E_y}{\partial x}, \frac{\partial H_z}{\partial t}, \frac{\partial H_z}{\partial x}$$

and hence Maxwell's equations become

$$\frac{\partial E_y}{\partial x} = -\frac{\mu}{c} \frac{\partial H_z}{\partial t} \tag{B.45}$$

and

$$-\frac{\partial H_z}{\partial x} = \frac{\varepsilon}{c} \frac{\partial E_y}{\partial t} \tag{B.46}$$

Taking the partial derivatives of Eqs. B.43 and B.44, we obtain

$$\frac{1}{v} E_y = -\frac{\mu}{c} H_z \qquad \text{and} \qquad \frac{1}{v} H_z = -\frac{\varepsilon}{c} E_y \tag{B.47}$$

which are equivalent since

$$\frac{c}{v} = \sqrt{\mu \varepsilon} \tag{B.48}$$

We then have

$$H_z = \sqrt{\frac{\varepsilon}{\mu}} E_y \tag{B.49}$$

Recalling that for most materials $\mu = 1$, we can generalize this to

$$|H| = \sqrt{\varepsilon}|E| \tag{B.50}$$

or

$$|H| = n|E| \tag{B.51}$$

Summarizing, the magnetic vector and the electric vector are perpendicular to each other and to the direction of propagation. The magnitude of the magnetic vector is equal to n times the magnitude of the electric vector, where n is the index of refraction of the medium.

B.4 REFERENCES

1. This introductory information was obtained from F. A. Jenkins and H. E. White, *Fundamentals of Optics*, McGraw-Hill, New York, 1957 p. 408.
2. Most of the information in this appendix was obtained from M. Born and E. Wolf, *Principles of Optics*, 4th edition, Pergamon, New York, 1969; and J. D. Jackson, *Classical Electrodynamics*, Wiley, New York, 1962.
3. R. M. A. Azzam and N. M. Bashara, *Ellipsometry and Polarized Light*, North Holland, Amsterdam, 1977.
4. This treatment is taken from M. Born and E. Wolf, *Principles of Optics*, 4th edition, Pergamon, New York, 1969, p. 612.

B.5 VECTOR CALCULUS NOTATION REVIEW

The textbook *Introduction to Vector Analysis* by H. F. Davis [Allyn and Bacon, Boston, 1961], provides very good physical descriptions and pictorial representations of the concepts of the vector calculus operations.

B.5.1 Vector Algebra

Suppose that $\bar{i}$, $\bar{j}$, and $\bar{k}$ are unit vectors in the x, y, and z directions, respectively, and that $\bar{U}$ and $\bar{V}$ are the vectors

$$\bar{U} = U_x\bar{i} + U_y\bar{j} + U_z\bar{k} \qquad \text{and} \qquad \bar{V} = V_x\bar{i} + V_y\bar{j} + V_z\bar{k}$$

The scalar product (or dot product) is

$$\bar{U} \cdot \bar{V} \equiv U_x V_x + U_y V_y + U_z V_z$$

The vector product (or cross product) is

$$\bar{U} \times \bar{V} \equiv (U_y V_z - U_z V_y)\bar{i} + (U_z V_x - U_x V_z)\bar{j} + (U_x V_y - U_y V_x)\bar{k}$$

B.5.2 Scalar and Vector Fields

Before getting into the vector calculus, we introduce the concepts of the scalar field and the vector field. If at each point in a region of space there is made to correspond a number (a scalar), then we say that the correspondence is a scalar field. Symbolically, for each point (x, y, z) there is a corresponding function $f(x, y, z)$. A scalar field is a scalar-valued function in three dimensions. Explicitly, the coordinates of a location is the input and a scalar value is the output.

Similarly, a vector field is when, for each point in space, there corresponds a vector. Symbolically, for the point (x, y, z) there corresponds the vector $\bar{F}(x, y, z)$. Explicitly, the input is the coordinates of a location and the output is a vector. In visualizing a vector field, we imagine that from each point in the field, there extends a vector. Both direction and magnitude may vary with position.

Examples of scalar-valued functions are temperature as a function of distance from one corner of a room, and atmospheric density measured north, east, and up from the Statue of Liberty. Examples of vector valued functions are the direction and intensity of the wind as a function of distance from one corner of a cornfield, and the electric field in the presence of several charged electrodes as a function of distance from some chosen reference.

B.5.3 Vector Calculus

We define the vector operator del to be

$$\nabla \equiv \bar{i}\frac{\partial}{\partial x} + \bar{j}\frac{\partial}{\partial y} + \bar{k}\frac{\partial}{\partial z}$$

Suppose that $f(x, y, z)$ is a scalar-valued function of x, y, and z. If we apply the vector operator described above to a scalar function, we obtain a vector function which is called the *gradient* of f:

$$\text{grad } f(x, y, z) \equiv \nabla f(x, y, z)$$

and is given by

$$\nabla f(x, y, z) \equiv \bar{i}\frac{\partial f}{\partial x} + \bar{j}\frac{\partial f}{\partial y} + \bar{k}\frac{\partial f}{\partial z}$$

The gradient is a vector which describes the rate of change of $f(x, y, z)$. Since the ∇f is a vector, it is reasonable to ask in which direction ∇f points. The scalar component of ∇f in any direction is the rate of change of f in that direction. Specifically, ∇f points in the direction of the maximum rate of change, and the magnitude of ∇f is the maximum rate of change of f per unit distance.

We have shown how the del operator is applied to a scalar-valued function to obtain a vector. There are two ways that this vector operator can be applied to a vector-valued function. One gives a scalar-valued function as a result and the other gives a vector-valued function.

Suppose $\bar{Q}(x, y, z)$ is a vector which is a function of position. We write $\bar{Q}$ as

$$\bar{Q} = Q_x\bar{i} + Q_y\bar{j} + Q_z\bar{k}$$

where the components Q_x, Q_y, and Q_z may be scalar functions of x, y, z, and t.

The operation which gives a scalar-valued function is the *divergence* represented as

$$\operatorname{div} Q \equiv \nabla \cdot \bar{Q}$$

and is given by

$$\nabla \cdot \bar{Q} \equiv \frac{\partial Q_x}{\partial x} + \frac{\partial Q_y}{\partial y} + \frac{\partial Q_z}{\partial z}$$

The divergence is a scalar function just as the dot product of two vectors is a scalar. Roughly speaking, the divergence tells us, at each point, the extent to which the field diverges away from the point.

A reasonable mental picture of the divergence is to consider a compressible fluid such as a gas, as suggested in Figure B.1. Suppose that the gas is moving in a chosen direction (usually as a function of time). In the figure, we show two unit volumes, one after the other. If the number of gas molecules passing through the second unit volume is lower than the number of gas molecules passing through the first unit volume, then the vector field is diverging. Note that divergence can be caused by a change in direction of the vectors with no change in magnitude, by a change in magnitude with no change in direction, or by a combination of both. Also note that the divergence is a scalar.

The operation which gives a vector-valued function is the *curl*.

$$\operatorname{curl} \bar{Q} \equiv \nabla \times \bar{Q}$$

that is,

$$\nabla \times \bar{Q} \equiv \bar{i}\left(\frac{\partial Q_z}{\partial y} - \frac{\partial Q_y}{\partial z}\right) + \bar{j}\left(\frac{\partial Q_x}{\partial z} - \frac{\partial Q_z}{\partial x}\right) + \bar{k}\left(\frac{\partial Q_y}{\partial x} - \frac{\partial Q_x}{\partial y}\right)$$

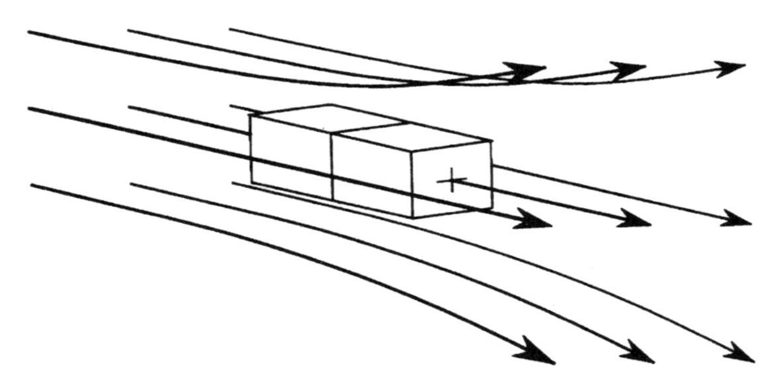

Figure B.1 The concept of divergence, shown heuristically.

Roughly speaking, the curl gives us, at each point, an indication of how the field swirls in the vicinity of that point. A reasonable mental picture is to consider a noncompressible fluid such as a liquid. Let us imagine that we have a paddle wheel in the fluid which is free to rotate about its axis, as suggested by Figure B.2. The angular velocity will depend on the orientation of the axis in the fluid. The curl is a vector, and its direction is along the axis of the paddle wheel when the paddle wheel is oriented so that its angular rotation is maximized.

The *Laplacian* operator is defined as

$$\nabla^2 = \frac{\partial^2}{\partial x^2} + \frac{\partial^2}{\partial y^2} + \frac{\partial^2}{\partial z^2}$$

The Laplacian of a function can be either a vector field or a scalar field, depending on the function. If $f(x, y, z)$ is a scalar-valued function, then the scalar Laplacian is

$$\nabla^2 f = \frac{\partial^2 f(x, y, z)}{\partial x^2} + \frac{\partial^2 f(x, y, z)}{\partial y^2} + \frac{\partial^2 f(x, y, z)}{\partial z^2}$$

Figure B.2 The concept of curl, shown heuristically.

If $\bar{Q}$ is a vector-valued function defined as

$$\bar{Q} = \bar{i}Q_x(x, y, z) + \bar{j}Q_y(x, y, z) + \bar{k}Q_z(x, y, z)$$

Then the vector Laplacian is

$$\nabla^2 \bar{Q} = \frac{\partial^2 \bar{Q}}{\partial x^2} + \frac{\partial^2 \bar{Q}}{\partial y^2} + \frac{\partial^2 \bar{Q}}{\partial z^2}$$

Note that the Laplacian is a 3-D counterpart of the ordinary second derivative.

Finally, we include a useful identity which can be verified by simply writing out the individual parts. The identity is

$$\nabla \times (\nabla \times \bar{Q}) \equiv \nabla(\nabla \cdot \bar{Q}) - \nabla^2 \bar{Q}$$

APPENDIX C

Snell's Law, Fresnel's Equations, and the Total Reflection Coefficient: Derivations and Historical Perspective

C.1 INTRODUCTION

The derivation of the equations of Snell and Fresnel will involve the wave equation rather significantly. We shall use the wave equation in its exponential form, and will find that Snell's law will come from the phase, that is, the exponent part, whereas Fresnel's equations will use the amplitude part. Both derivations will involve external conditions or boundary conditions which come from the physics of materials. The equation for dealing with two interfaces will involve setting up the problem properly and then some clever algebraic manipulations.

Along with the derivations, we will also mention some historical[1] aspects to put into perspective the time of the original derivations.

C.2 REFLECTION AND REFRACTION (SNELL'S LAW)

The stereotypical drawing for reflection and refraction from a single interface is shown in Figure C.1. Light is incident onto a dielectric material at an angle of incidence (measured from the normal) of ϕ_i. The reflected light makes an angle ϕ_r to the normal, and the refracted light is transmitted in the dielectric material at an angle ϕ_t.

Of all of the laws and equations in optics, the law of reflection is probably the most obvious, and is almost trivial. Among the ancient Greeks, Euclid, in about 300 BC, enunciated the fact that the angle of incidence, ϕ_i, and the angle of reflection, ϕ_r, are equal. The law of refraction is not so obvious. Ptolemy of Alexandria (70–147 AD) made measurements at the air–water, air–glass, and glass–water interfaces and concluded that the ratio of the angles of incidence

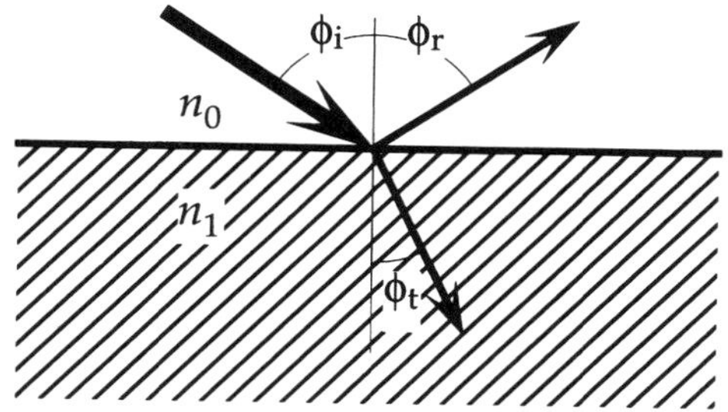

Figure C.1 Light reflecting from and passing through an interface between an ambient characterized by index n_0 and a material characterized by index n_1.

and refraction remained constant for various angles of incidence at a given interface. We know now that this is only an approximation, and that it is the ratio of the sines of the angles which remains constant. Because of the developing attitude at that time, that knowledge was to be obtained by revelation rather than by experimental observation, it was roughly one and a half millennia later before additional progress was made in this matter.

At the beginning of the seventeenth century there was a rebirth of scientific activity, and in 1621, Willebrord Snell, Professor of Mathematics at Leyden (Leiden, in The Netherlands), discovered experimentally what we now call Snell's law. Snell died about 5 years later, never having made his work public. Rene Descartes later published this work, and it is generally believed that Descartes had seen Snell's manuscript on this subject. This appears to be Snell's one and only lasting contribution to science.

To put this in perspective, Snell and Descartes were contemporaries of Galileo. The Inquisition in Europe was just winding down. On the Atlantic side of the New World, the *Mayflower* had just landed at Plymouth Rock. In the western part of the New World, Don Juan de Oñate of Mexico had established a provincial government in New Mexico, but no established civilization was yet present in Arizona. It was shortly before this time that Oñate, on a trip west from New Mexico, inscribed his name on a large rock formation which is now El Morro National Monument in northwestern New Mexico. On this trip, he also had discovered that California was not an island, but was attached to the continent.

In Appendix B, we showed that the equation for electromagnetic waves consisted of an amplitude and an exponential term which contained the time and space variation of the wave. For a transparent medium, the equation is

$$\bar{E}(\bar{r}, t) = \bar{E}^0 \, e^{-j(2\pi/\lambda)\bar{r}\cdot\bar{s}} e^{j\omega t} \tag{C.1}$$

where $\bar{r}$ is the position vector of a location (x, y, z) in space relative to an origin, $\bar{s}$ is a unit vector in the direction of propagation, $\bar{E}^0$ is the amplitude, and λ is the wavelength in the medium of interest.

We now rewrite[2] Eq. C.1 as

$$\bar{E}(\bar{r}, t) = \bar{E}^0 \, e^{j\omega t - j(2\pi/\lambda)\bar{r}\cdot\bar{s}} \tag{C.2}$$

Recalling that

$$\frac{2\pi}{\lambda} = \frac{\omega}{v}$$

where v is the velocity in the medium of interest, we then write Eq. C.2 as

$$\bar{E}(\bar{r}, t) = \bar{E}^0 e^{j\omega(t - \bar{r}\cdot\bar{s}/v)} \tag{C.3}$$

We now use a rectilinear coordinate system, as suggested in Figure C.2, with the x axis along the surface and the z axis perpendicular to the surface. The y axis is perpendicular to the page. The position vector $\bar{r}$ can be written as $\bar{r} = \bar{i}x + \bar{j}y + \bar{k}z$, where $\bar{i}$, $\bar{j}$, and $\bar{k}$ are unit vectors in the x, y, and z directions, respectively.

In our illustrations, we often draw rays, as shown in Figure C.1. We must keep in mind, however, that the light arrives at the surface as a plane wave, as suggested in Figure C.2. A plane wave propagated in the direction specified by the unit vector is completely determined in space by Eq. C.3 when the time variation is known at a particular point (which we choose to call the origin of

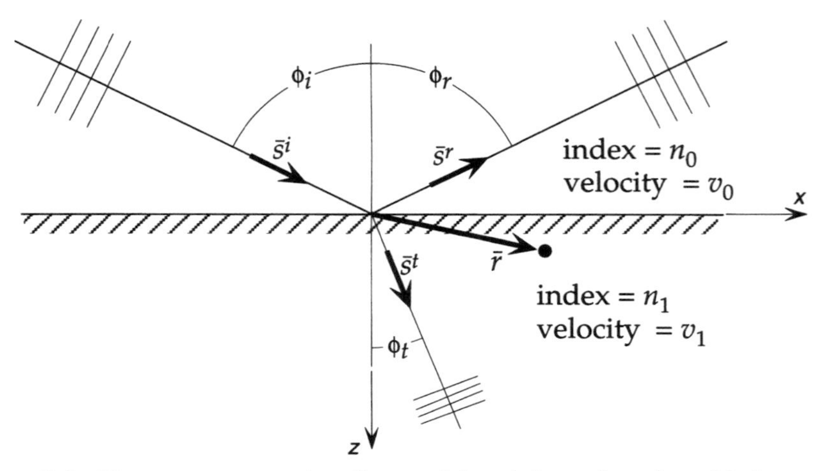

Figure C.2 Plane waves at an interface: arriving, being reflected, and being transmitted. $\bar{r}$ is a position vector of a point of interest, and $\bar{s}^{\mathrm{i}}$, $\bar{s}^{\mathrm{r}}$, and $\bar{s}^{\mathrm{t}}$ are unit vectors in the direction of propagation of the incident, reflected, and transmitted wave, respectively. The x and z directions are indicated. The y direction is perpendicular to the plane of the paper.

our coordinate system). The same conditions which exist at the origin at any given time t exist at point $\bar{r}$ at a time interval $\bar{r} \cdot \bar{s}/v$ later.

Let $\bar{s}^{\mathrm{i}}$ be the unit vector for the incident wave, $\bar{s}^{\mathrm{r}}$ be the unit vector for the reflected wave, and $\bar{s}^{\mathrm{t}}$ be the unit vector for the transmitted wave. We denote the velocity in the ambient as v_0 and the velocity in the refracting medium as v_1. We can write

$$\bar{s}^{\mathrm{i}} = \bar{i} \sin \phi_{\mathrm{i}} + \bar{k} \cos \phi_{\mathrm{i}}$$

The first requirement of continuity for the three (incident, reflected, and transmitted) electric vectors and their corresponding magnetic vectors is that they satisfy the boundary conditions at all times. This yields the result that the frequency ω is the same for the incident, reflected, and transmitted wave. We use the term ω for all three rather than ω_{i}, ω_{r}, and ω_{t}.

On the boundary between the two media, the argument of the exponential for the incident wave must be equal to the argument of the exponential for the reflected wave, and also equal to the argument for the transmitted wave. This is to say that for $\bar{r}$ on the boundary,

$$t - \frac{\bar{r} \cdot \bar{s}^{\mathrm{i}}}{v_0} = t - \frac{\bar{r} \cdot \bar{s}^{\mathrm{r}}}{v_0} = t - \frac{\bar{r} \cdot \bar{s}^{\mathrm{t}}}{v_1}$$

or simply

$$\frac{\bar{r} \cdot \bar{s}^{\mathrm{i}}}{v_0} = \frac{\bar{r} \cdot \bar{s}^{\mathrm{r}}}{v_0} = \frac{\bar{r} \cdot \bar{s}^{\mathrm{t}}}{v_1} \tag{C.4}$$

This equation can also be derived from Huygens' principle, which requires that the incident wave and the reflected and transmitted waves (which can be considered as wavelets generated by the incident wave) have the same frequency and phase as the incident wave.

The dot product of the position vector with the unit vector for the incident wave is

$$\bar{r} \cdot \bar{s}^{\mathrm{i}} = x \sin \phi_{\mathrm{i}} + z \cos \phi_{\mathrm{i}}$$

and for points on the interface where $z = 0$, we have

$$\bar{r} \cdot \bar{s}^{\mathrm{i}} = x \sin \phi_{\mathrm{i}}$$

and similarly for the reflected and transmitted wave. Equation C.4 then gives

$$\frac{x \sin \phi_{\mathrm{i}}}{v_0} = \frac{x \sin \phi_{\mathrm{r}}}{v_0} = \frac{x \sin \phi_{\mathrm{t}}}{v_1} \tag{C.5}$$

The law of reflection comes from the first two terms, which imply that the sine of the angle of incidence is equal to the sine of the angle of reflection.

There is a subtlety here based on how we choose to draw our angles which will be important for later derivations. Because of the dot product between the position vector and the unit vectors, there is an ambiquity between signs. The above arguments imply that either $\phi_r = \phi_i$ or $\phi_r = \pi - \phi_i$. The latter is true, since the former would imply that the reflected ray would travel back the same direction that the incident ray came from. This implies that $\cos\phi_r = -\cos\phi_i$.

We use the first and third terms to give

$$\frac{\sin\phi_i}{v_0} = \frac{\sin\phi_t}{v_1}$$

The index of refraction n is defined as $n = c/v$, where c is the velocity of light in a vacuum. With this we get

$$n_0 \sin\phi_i = n_1 \sin\phi_t \tag{C.6}$$

which is Snell's Law.

This derivation can be summarized as follows. Consider the wavefront as it propagates along the interface. Since the velocity is different in the two media, the angle of the direction of propagation (for nonnormal incidence) must adjust so that the wavefront immediately above the interface is continuous with (does not get ahead of or lag behind) the wavefront immediately below the interface.

Drude[3] points out that

"The laws of reflection and refraction follow, then, from the fact of the existence of boundary conditions and are altogether independent of the particular form of these conditions."

C.3 FRESNEL'S REFLECTION AND TRANSMISSION COEFFICIENTS

In our historical perspective, we jump ahead two centuries to the early 1800s. In the intervening time, the works of Huygens, Hooke, Newton, and others have significantly increased the scientific understanding of light. The unequivocal establishment of the wave nature of light came at the beginning of the nineteenth century with the discovery of polarized light by Malus, and the work of Young in Scotland and Fresnel in France.

There was some difficulty with the concept of *influence at a distance*, and the burning question for the wave nature of light was "what was waving?"

Huygens had postulated an "aether" which pervades all space as the medium through which this wave traveled. Fresnel developed a theory from which came the Fresnel reflection and transmission coefficients. In his theory, he postulated the necessary properties for the aether to give the required wave properties. "His method was to work backwards from the known properties of light, in the hope of arriving at a mechanism to which they could be attributed..."[4] Fresnel made numerous contributions to the development of the scientific understanding of light. The coefficients are a small part of Fresnel's total contribution.

To put the time frame into perspective, Lewis and Clark had finished their westward exploration and the United States had just finished the second war with England (The War of 1812). California was a thriving colony of Mexico and the Spanish rule of Mexico was about to come to an end.

The derivation of the Fresnel reflection coefficients[5] from Maxwell's equations depends on the requirement that there be no discontinuity of the tangential components of the electric field and the magnetic field at the interface. As indicated earlier, we will use the amplitude for this derivation rather than the exponent, which was the case for Snell's law.

In Figure C.3, we again take the plane of incidence as the x–z plane with the y direction going out of the page. We separate the electric field vector into components in the plane of incidence (the p-wave) and perpendicular to the plane of incidence (the s-wave). In this figure, we show the amplitude of electric vectors in the plane of incidence (the p-wave) only.

The amplitude of the incident wave is denoted as $E_{\mathrm{i}}^{\mathrm{p}}$, the amplitude of the reflected wave is denoted as $E_{\mathrm{r}}^{\mathrm{p}}$, and the amplitude of the transmitted (or

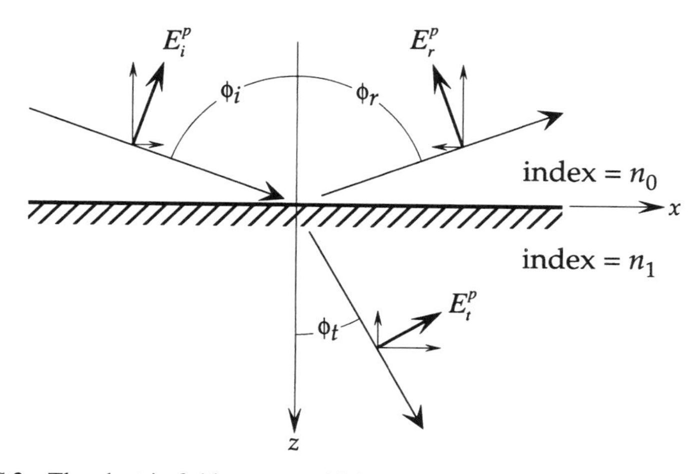

Figure C.3 The electric field vectors which are used in applying the boundary conditions. We show only the components in the plane of incidence (the x–z plane). The component which is perpendicular to the plane of incidence (in the y direction) is not shown.

refracted) wave is denoted as E_t^p. E_i^s, E_r^s, and E_t^s are the corresponding amplitudes perpendicular to the plane of incidence (not shown). We also further break down the p-wave electric vector into its components which are parallel to the surface and perpendicular to the surface. The components which are parallel to the surface in the x direction are denoted as E_i^x, E_r^x, and E_t^x. The corresponding quantities for the magnetic wave are denoted as H_i^p, H_r^p, H_t^p, H_i^s, H_r^s, H_t^s, H_i^x, H_r^x, and H_t^x.

The quantities which we seek are the Fresnel reflection coefficients

$$r^p = \frac{E_r^p}{E_i^p} \quad \text{and} \quad r^s = \frac{E_r^s}{E_i^s} \tag{C.7}$$

and the Fresnel transmission coefficients

$$t^p = \frac{E_t^p}{E_i^p} \quad \text{and} \quad t^s = \frac{E_t^s}{E_i^s} \tag{C.8}$$

The boundary condition which must be satisfied is that the sum of the tangential components of the electric field just above the interface must equal the sum of the components just below the interface. This condition must also be satisfied for the magnetic field components. In equation form, the boundary conditions become

$$E_i^x + E_r^x = E_t^x \tag{C.9}$$

$$E_i^y + E_r^y = E_t^y \tag{C.10}$$

$$H_i^x + H_r^x = H_t^x \tag{C.11}$$

$$H_i^y + H_r^y = H_t^y \tag{C.12}$$

With substitution, the left-hand side of Eq. (C.9) becomes

$$E_i^p \cos \phi_i + E_r^p \cos \phi_r = (E_i^p - E_r^p) \cos \phi_i$$

since $\cos \phi_r = -\cos \phi_i$. There is only one tangential component below the interface, that is, $E_t^p \cos \phi_t$, and hence Eq. C.9 becomes

$$(E_i^p - E_r^p) \cos \phi_i = E_t^p \cos \phi_t \tag{C.13}$$

For the s-wave, Eq. C.10 gives us simply

$$E_i^s + E_r^s = E_t^s \tag{C.14}$$

Similarly for the magnetic components, Eq. C.11 becomes

$$(H_i^p - H_r^p)\cos\phi_i = H_t^p \cos\phi_t \tag{C.15}$$

and Eq. C.12 becomes

$$H_i^s + H_r^s = H_t^s \tag{C.16}$$

The magnetic field is not independent of the electric field (see Eq. B.49). In general, in vector format, $\bar{H} = n\bar{s} \times \bar{E}$ where, as before, $\bar{s}$ is the unit vector in the direction of propagation. This allows us to write Eqs. (C.15 and C.16) in terms of the electric field rather than the magnetic field, and our boundary conditions become

$$(E_i^p - E_r^p)\cos\phi_i = E_t^p \cos\phi_t \tag{C.17}$$

$$E_i^s + E_r^s = E_t^s \tag{C.18}$$

$$n_0(E_i^s - E_r^s)\cos\phi_i = n_1 E_t^s \cos\phi_t \tag{C.19}$$

$$n_0(E_i^p + E_r^p) = n_1 E_t^p \tag{C.20}$$

Note that two of the equations deal with components parallel to the plane of incidence and two deal with components perpendicular to the plane of incidence. Using the definitions of the Fresnel coefficients defined in Eqs. C.7 and C.8, the above four equations can be solved algebraically to give

$$t^p = \frac{2n_0 \cos\phi_i}{n_1 \cos\phi_i + n_0 \cos\phi_t}$$

$$t^s = \frac{2n_0 \cos\phi_i}{n_0 \cos\phi_i + n_1 \cos\phi_t}$$

$$r^p = \frac{n_1 \cos\phi_i - n_0 \cos\phi_t}{n_1 \cos\phi_i + n_0 \cos\phi_t}$$

$$r^s = \frac{n_0 \cos\phi_i - n_1 \cos\phi_t}{n_0 \cos\phi_i + n_1 \cos\phi_t}$$

More generally, for light in medium a approaching an interface between medium a and medium b, the Fresnel coefficients are

$$t_{ab}^p = \frac{2n_a \cos\phi_a}{n_b \cos\phi_a + n_a \cos\phi_b} \tag{C.21}$$

$$t_{ab}^{s} = \frac{2n_a \cos \phi_a}{n_a \cos \phi_a + n_b \cos \phi_b} \tag{C.22}$$

$$r_{ab}^{p} = \frac{n_b \cos \phi_a - n_a \cos \phi_b}{n_b \cos \phi_a + n_a \cos \phi_b} \tag{C.23}$$

$$r_{ab}^{s} = \frac{n_a \cos \phi_a - n_b \cos \phi_b}{n_a \cos \phi_a + n_b \cos \phi_b} \tag{C.24}$$

Note that if the direction of propagation is reversed, that is, the light approaches the interface between medium a and medium b from medium b, then the Fresnel reflection coefficients can readily be shown (for either the p-waves or the s-waves) to be

$$r_{ba} = -r_{ab} \tag{C.25}$$

$$t_{ba} = (1 - r_{ab}^2)/t_{ab} \tag{C.26}$$

C.4 USING THE COMPLEX INDEX OF REFRACTION

Snell's law and the Fresnel coefficients were undoubtedly originally developed for dielectrics, and hence in the above derivations we have used the lower case letter for the index of refraction, implying a real number. We state (without proof) that these equations (C.6 and C.21–C.24) are still valid if one uses the complex index of refraction, $\tilde{N}$, where $\tilde{N} = n - jk$.

The most general form of Snell's law, is then

$$\tilde{N}_a \sin \phi_a = \tilde{N}_b \sin \phi_b \tag{C.27}$$

Suppose medium a is the ambient and has a zero extinction coefficient, that is, $\tilde{N}_a$ and $\sin \phi_a$ are both real numbers. If medium b has a nonzero extinction coefficient, that is, $\tilde{N}_b$ has an imaginary component, then the function $\sin \phi_b$ is a complex function rather than the usual real trigonometric function. There is a corresponding complex function $\cos \phi_b$, and the relationship between these two functions is

$$\sin^2 \phi_b + \cos^2 \phi_b = 1$$

The more general Fresnel coefficients are given by

$$r_{ab}^{p} = \frac{\tilde{N}_b \cos \phi_a - \tilde{N}_a \cos \phi_b}{\tilde{N}_b \cos \phi_a + \tilde{N}_a \cos \phi_b} \tag{C.28}$$

$$r^{s}_{ab} = \frac{\tilde{N}_a \cos\phi_a - \tilde{N}_b \cos\phi_b}{\tilde{N}_a \cos\phi_a + \tilde{N}_b \cos\phi_b} \tag{C.29}$$

$$t^{p}_{ab} = \frac{2\tilde{N}_a \cos\phi_a}{\tilde{N}_b \cos\phi_a + \tilde{N}_a \cos\phi_b} \tag{C.30}$$

$$t^{s}_{ab} = \frac{2\tilde{N}_a \cos\phi_a}{\tilde{N}_a \cos\phi_a + \tilde{N}_b \cos\phi_b} \tag{C.31}$$

C.5 TOTAL REFLECTION COEFFICIENT FOR A FILM ON A SUBSTRATE

The total reflection coefficient is the ratio of the *amplitude* of the total reflected wave to the *amplitude* of the incident wave, and is denoted as either R^p or R^s for the p-waves or s-waves, respectively. The equation for R^p will involve the Fresnel reflection and transmission coefficients for the p-waves, and that for R^s will involve the Fresnel reflection and transmission coefficients for the s-waves. The derivation is the same for both R^p and R^s. Accordingly, we omit the superscript which denotes either the s-wave or the p-wave and restore it at the end of the derivation.

For an interface between an ambient such as air and a bulk material, the ratio of the amplitude of the reflected light to the incident light can be calculated from the above derived Fresnel reflection coefficients. The situation is more complex when a film is present, however, owing to the presence of a second interface, as suggested by Figure C.4.

There is some disagreement as to the historical origin of this derivation. Azzam and Bashara[6] attribute it to Drude in 1890, whereas Born and Wolf[7] attribute it to Airy in 1833 ("derived in a different manner"). The earlier figure would put this at about the same time as Fresnel. Drude is considered to be the pioneer of ellipsometry, and is generally given credit for the derivation. Vasicek[8] puts this disagreement in perspective and suggests that the formula for the total reflection coefficient be called the "Airy-Drude formulae".

We suppose[9] that the light is incident from the ambient (material 1) onto the interface between material 1 and material 2. For convenience, let us take the amplitude of the incident light to be unity. The amplitude of the first reflection (indicated as partial wave 1) will then be r_{12}, and the amplitude of the transmitted light in medium 2 after this first interaction will be t_{12}.

If only one interface were present, the transmitted light would continue and never have to be dealt with again. When a second interface is present, some of this light is transmitted into medium 3 and some is reflected back into medium 2, headed back toward the upper interface. The amplitude of the returning partial wave is $t_{12}r_{23}$. When this beam interacts with the upper interface,

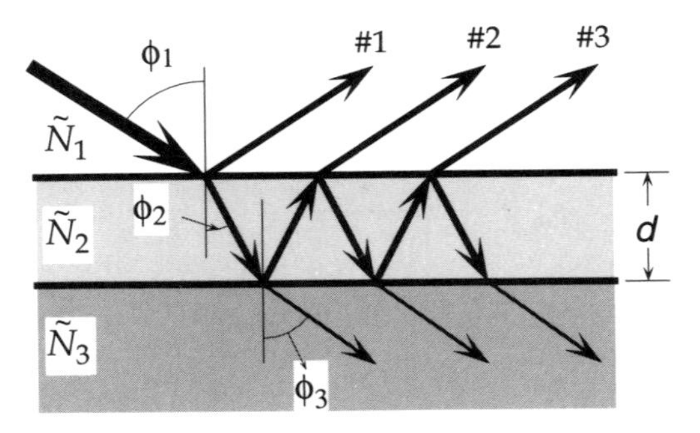

Figure C.4 Light reflecting from a single film on a substrate. Several of the partial waves are shown. The total reflected wave is the sum of the partial waves going back into the ambient.

some is transmitted into medium 1 and some is reflected, headed again toward the lower interface. The amplitude of the part transmitted into medium 1 (partial wave 2) will then be $t_{12}r_{23}t_{21}$. This second partial wave will be out of phase with the first partial wave and the complex amplitude must include the phase factor $e^{-j2\beta}$ where β is the phase change that the wave experiences as it traverses the film once between interfaces. The phase angle β, or the film phase thickness, is given by

$$\beta = 2\pi\left(\frac{d}{\lambda}\right)\tilde{N}_2 \cos \phi_2 \tag{C.32}$$

The complex amplitudes of the successive partial waves are then

$$r_{12}$$

$$t_{12}r_{23}t_{21}e^{-j2\beta}$$

$$t_{12}r_{23}r_{21}r_{23}t_{21}e^{-j4\beta}$$

$$t_{12}r_{23}r_{21}r_{23}r_{21}r_{23}t_{21}e^{-j6\beta}$$

and so on.

Addition of the partial waves leads to an infinite geometric series for the total reflected amplitude. After rearranging some of the terms, we have the series

$$R = r_{12} + t_{12}t_{21}r_{23}e^{-j2\beta} + t_{12}t_{21}r_{21}r_{23}^2e^{-j4\beta} + t_{12}t_{21}r_{21}^2r_{23}^3e^{-j6\beta} + \cdots$$

which can be expressed as

$$R=r_{12}+t_{12}t_{21}r_{23}e^{-j2\beta}[1+(r_{21}r_{23}e^{-j2\beta})^1+(r_{21}r_{23}e^{-j2\beta})^2+(r_{21}r_{23}e^{-j2\beta})^3+\cdots]$$

Using the relationship

$$\frac{1}{1-x}=1+x+x^2+x^3+x^4+\cdots \qquad (x^2<1)$$

we can write

$$R=r_{12}+\frac{t_{12}t_{21}r_{23}e^{-j2\beta}}{1-r_{21}r_{23}e^{-j2\beta}}$$

Using Eq. C.25 and C.26, we obtain the total reflection coefficient

$$R=\frac{r_{12}+r_{23}e^{-j2\beta}}{1+r_{12}r_{23}e^{-j2\beta}} \tag{C.33}$$

In a similar manner we can obtain the total transmission coefficient, which is given by

$$T=\frac{t_{12}t_{23}e^{-j\beta}}{1+r_{12}r_{23}e^{-j2\beta}} \tag{C.34}$$

As stated earlier, the derivation is the same for s-waves as for p-waves. We now restore the polarization information by writing these as

$$R^{\mathrm{p}}=\frac{r_{12}^{\mathrm{p}}+r_{23}^{\mathrm{p}}e^{-j2\beta}}{1+r_{12}^{\mathrm{p}}r_{23}^{\mathrm{p}}e^{-j2\beta}} \tag{C.35}$$

$$R^{\mathrm{s}}=\frac{r_{12}^{\mathrm{s}}+r_{23}^{\mathrm{s}}e^{-j2\beta}}{1+r_{12}^{\mathrm{s}}r_{23}^{\mathrm{s}}e^{-j2\beta}} \tag{C.36}$$

$$T^{\mathrm{p}}=\frac{t_{12}^{\mathrm{p}}t_{23}^{\mathrm{p}}e^{-j\beta}}{1+r_{12}^{\mathrm{p}}r_{23}^{\mathrm{p}}e^{-j2\beta}} \tag{C.37}$$

$$T^{\mathrm{s}}=\frac{t_{12}^{\mathrm{s}}t_{23}^{\mathrm{s}}e^{-j\beta}}{1+r_{12}^{\mathrm{s}}r_{23}^{\mathrm{s}}e^{-j2\beta}} \tag{C.38}$$

These are the total reflection coefficients and the total transmission coefficients for a film-covered surface.

It is interesting to note that Eqs. C.35 and C.36 can be used to calculate the total reflection coefficients of a stack with any number of layers. Consider two layers on a substrate. First calculate the total reflection coefficient R (p- or s-) for the bottom film on the substrate as if the top film did not exist. Then calculate the final total reflection coefficient from Eq. C.35 or C.36 by inserting the total R from the previous step as r_{23}. This can be repeated for any number of layers, is easy to implement, and is computationally efficient. McCrackin[10] used this method in his computer program in 1969 and the method was also used for calculation in an appendix by Tompkins.[11]

C.6 REFERENCES

1. The sources of historical material for this appendix are: M. Born and E. Wolf, *Principles of Optics*, 4th edition, Pergamon, New York, 1969; J. Strong, *Concepts of Classical Optics*, Freeman, San Francisco, 1958; Sir Edmund Whittaker, *A History of the Theories of Aether and Electricity: I. The Classical Theories*, Tomash/American Institute of Physics, New York, first published 1910, revised 1951; Ernst Mach, *The Principles of Physical Optics: An Historical and Philosophical Treatment*, Dover Publications, New York, 1926.
2. The derivation of Snell's law is patterned after that in M. Born and E. Wolf, *Principles of Optics*, 4th edition, Pergamon, New York, 1969, p. 37.
3. P. Drude, *The Theory of Optics*, Longmans, Green, New York, 1902, p. 282.
4. Sir Edmund Whittaker, *A History of the Theories of Aether and Electricity: I. The Classical Theories*, Tomash/American Institute of Physics, New York, first published 1910, revised 1951, p. 125.
5. The derivation of Fresnel's equations is patterned after that in M. Born and E. Wolf, *Principles of Optics*, 4th edition, Pergamon, New York, 1969, p. 38 and following.
6. R. M. A. Azzam and N. M. Bashara, *Ellipsometry and Polarized Light*, North-Holland, Amsterdam, 1977, p. 284. Their reference is to P. Drude, *Ann. Phys. Chem.*, **39**, 481 (1890). This is incorrect, as the correct reference should be P. Drude, *Ann. Phys. Chem.*, **36**, 865 (1889).
7. M. Born and E. Wolf, *Principles of Optics*, 4th edition, Pergamon, New York, 1969, p. 62.
8. A. Vasicek, *Appl. Opt.*, **4**, 1032 (1965).
9. The derivation of the total reflection and transmission coefficients is patterned after that in R. M. A. Azzam and N. M. Bashara, *Ellipsometry and Polarized Light*, North-Holland, Amsterdam 1977, p. 283 and following.
10. F. L. McCrackin, Nat. Bur. Stand., Tech. Note 479, 1969.
11. H. G. Tompkins, *A User's Guide to Ellipsometry*, Academic Press, New York, 1993, Appendix A.

INDEX